Empower Decision Makers with SAP Analytics Cloud

Modernize BI with SAP's Single Platform for Analytics

Vinayak Gole
Shreekant Shiralkar

Apress®

Empower Decision Makers with SAP Analytics Cloud: Modernize BI with SAP's Single Platform for Analytics

Vinayak Gole
Mumbai, India

Shreekant Shiralkar
Mumbai, Maharashtra, India

ISBN-13 (pbk): 978-1-4842-6096-8
https://doi.org/10.1007/978-1-4842-6097-5

ISBN-13 (electronic): 978-1-4842-6097-5

Managing Director, Apress Media LLC: Welmoed Spahr
Acquisitions Editor: Divya Modi
Development Editor: Matthew Moodie
Coordinating Editor: Divya Modi

Cover designed by eStudioCalamar

Cover image designed by Pixabay

Distributed to the book trade worldwide by Springer Science+Business Media New York, 1 New York Plaza, New York, NY 10004. Phone 1-800-SPRINGER, fax (201) 348-4505, e-mail orders-ny@springer-sbm.com, or visit www.springeronline.com. Apress Media, LLC is a California LLC and the sole member (owner) is Springer Science + Business Media Finance Inc (SSBM Finance Inc). SSBM Finance Inc is a **Delaware** corporation.

For information on translations, please e-mail booktranslations@springernature.com; for reprint, paperback, or audio rights, please e-mail bookpermissions@springernature.com.

Apress titles may be purchased in bulk for academic, corporate, or promotional use. eBook versions and licenses are also available for most titles. For more information, reference our Print and eBook Bulk Sales web page at http://www.apress.com/bulk-sales.

Any source code or other supplementary material referenced by the author in this book is available to readers on GitHub via the book's product page, located at www.apress.com/978-1-4842-6096-8. For more detailed information, please visit http://www.apress.com/source-code.

Printed on acid-free paper

To my wife Shweta and son Vivaan, who have been the pillars of strength in my life; and my parents who have been my guiding stars. This book would not have been possible without their support.

—Vinayak

To all my co-authors and contributors.

Mentoring authors and helping coworkers realize their aspirations have led me on an extremely satisfying and rewarding journey. This book celebrates one more milestone in such a journey and has been additionally rewarding as it engaged me productively in the time of COVID-19 that otherwise offered extremely limited avenues of mindfulness. I therefore thank Vinayak Gole for allowing me to be his coach and mentor in accomplishing his aspiration that began with writing a few technical articles and now fructified into this book.

I take this opportunity to express my gratitude to each of my co-authors on technical articles and books. I begin with Bharat Patel who partnered in conceptualizing and developing the first technical article. Recognition and reward from the article fueled the exciting journey in co-authoring a few of the most read and referred-to articles and later reaching the important milestone of authoring our first book.

Amol Palekar hopped on the journey soon, initially co-authoring a few of the most referred-to and popular technical articles and later creating one of the most cherished milestones on my journey of authoring, a bestselling book on SAP Analytics. The book and its later editions retained its high selling position for a very long time.

More recently, I felt accomplished to have coached and co-authored with Avijit Dutta and Deepak Sawant by conceiving and authoring books on technology in supply chain management and SAP Analytics.

The journey of mentoring has had many more moments and mentees who added important milestones, namely Abhijit Ghate, Sohil Shah, Meenakshi Chopra, Rohit Kumar Das, Achin Kimtee, and Jyoti Jain to name a few.

—Shreekant

Table of Contents

About the Authors

Vinayak Gole is a seasoned analytics consultant with experience across multiple business domains and roles. As senior architect at Tata Consultancy Services Ltd., Vinayak has been engaged in technology consulting and architecting solutions across the SAP Analytics Portfolio for Fortune 500 firms. He has been instrumental in building, mentoring, and enabling teams delivering complex digital transformations for global clients. Passionate about technology, Vinayak regularly publishes articles and technical papers with well-known publications. He is also an active contributor to the SAP community and regularly publishes blogs on technologies in the SAP Analytics Portfolio.

Shreekant Shiralkar is a senior management professional with expertise on leading and managing business functions and technology consulting. He established and developed business units for Fortune 500 firms, namely a public service business for the world's leading professional services company and launched the Shell Gas business in India for a JV of Shell. Shreekant grew the SAP technology business for Tata Consultancy Services Ltd. by winning strategic clients in new and existing geographies, creating innovative service offerings. He played a critical part in multiple transformation programs for Bharat Petroleum Corporation Ltd. He has mentored authors, published bestselling books and white papers on technology, and has patents on technology and service delivery. He specializes in realizing concepts to their value-creation stage, innovation and transformation, and building organizations.

About the Technical Reviewer

Atul Thatte is an Analytics expert having more than sixteen years of work experience. He has delivered several large end-to-end implementation projects in technology areas involving SAP BW, HANA, and CPM. For more than a year now, Atul has been working in the SAP Analytics Cloud area.

Acknowledgments

We would like to express our sincere gratitude to Atul Thatte, who reviewed and provided invaluable feedback on the contents of this book. Atul is a Business Intelligence professional with extensive experience across the entire SAP Analytics Portfolio.

We would also like to thank Divya Modi, Matthew Moodie, and the entire publication team at Apress for their unflinching support during the entire duration of this book from conceptualization to publication.

—Vinayak and Shreekant

Introduction

The COVID-19 pandemic has highlighted the uncertainties and rapidly changing environment: we are experiencing mostly uncharted waters, and tides continue to shift. It's therefore not surprising that analytics, widely recognized for its problem-solving and predictive prowess, is assumed to be the most essential lever to navigate successfully for every enterprise. Further, the time to comprehend and make decisions has been severely reduced and decision makers have to quickly learn and adopt analytics solutions. Analytics capabilities that once might have taken these enterprises months or years to build are now expected to be ready in a matter of weeks if not days. Enterprises are therefore increasingly embracing advanced analytic solutions that take the least efforts and time to become purposeful. Innovation in technology is moving enterprises to adopt and integrate technologies like cloud and augmented analytics to extract the best information rapidly by analyzing humongous amounts of data.

In the aforesaid circumstance, the content in the book is contextualized and deals with the subject by alluding to a typical multinational enterprise and the challenges faced, more specifically in its current analytics landscape and the incident expectations of a modern analytics solution. The book explores the organization's current drawbacks and requirements progressively through the chapters. The chapters explore each of these expectations and align them with capabilities of the SAP Analytics Cloud (SAC) along with possible solutions, customer benefits, and a future road map.

The contents of the book are designed for easy learning while aligning to the business problem and expectations at hand. Step-by-step processes aligned with illustrations and screenshots are provided to enable readers to simulate the business problem at hand. Simulating the problem and the provided step-by-step process to achieve the solution will enable readers to achieve maximum knowledge from the book.

The book begins by exploring the current trends in analytics and how SAP has built a strong portfolio of products to align with the trends. This is explored in Chapter 1.

Chapter 2 explores the current challenges faced by ABC Inc. in their current landscape and the expectations from the enterprise from a new, modern analytics landscape.

Chapters 3 through 8 explore each of these capacities in detail and how they are aligned to the requirements of ABC Inc. from the analytics landscape.

Chapter 9 explores how SAC enables a secure landscape for analytics requirements, which is a core expectation from any cloud-based application. Chapter 10 explains the future road map for SAC and the new capabilities expected.

The book follows a unique approach to understanding SAC as an application with respect to a business scenario. The book has been written to cater to the needs of a variety of users including business users and beginners who would like to understand self-service analytics and explore data analytics to the fullest.

CHAPTER 1

Current Trends in Analytics and SAP's Road Map

Running a successful enterprise depends on the ability to gain insight into enterprise data and to extract and present information in a meaningful way. A company's employees need to be able to transform data into actionable insights no matter where they are located in the enterprise.

Business Intelligence

Let us appreciate how and why Business Intelligence (BI) is essential for an Intelligent Enterprise. At the beginning of the century, while technology was proliferating across enterprise functions and its ubiquitous nature was fueling a data explosion, the availability of data with its variety at huge velocity gave birth to concepts like Big Data. Figure 1-1 is a reflection on how BI gained center stage.

© Vinayak Gole, Shreekant Shiralkar 2020
V. Gole and S. Shiralkar, *Empower Decision Makers with SAP Analytics Cloud*,
https://doi.org/10.1007/978-1-4842-6097-5_1

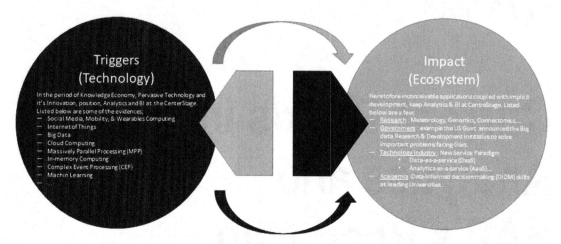

Figure 1-1. *Triggers and Impact: Analytics and Business Intelligence*

Technological invention brought down the cost of BI technology whether it was for sourcing the data or storage or processing the data, paving the way to the phenomenon and fact that "Data is the new oil." Alongside value derivation from data, data monetization started to gain attention and gave motivation to associated fields like data science and advanced analytics including predictive analytics. Figure 1-2 summarizes the value from BI.

Figure 1-2. *Value from BI*

Not long ago, analytics was a domain limited to IT and data analysts, who supported the decision-makers by delivering data visualizations. In the digital economy, self-service has become the norm that mandates software solutions to be intuitive and easily adopted and used by the regular business user across the enterprise; and there's a trend for the end user to learn the solution, understand the data models, and manage the content. Artificial intelligence technologies embedded within analytical applications have the potential to be able to make operational decisions without much human intervention. Analytical applications will augment end users to perform complex analytical tasks with algorithms, machine learning, and new natural language processing-based voice interfaces.

Applications and the use of analytics added developments to self-service, analytics specific to an industry like retail, and specific functions, for example, advertising. Planning functions, due to their similarity with technical uses and applications, started getting integrated with analytic solutions and offerings. Lastly but equally important, business changes were needed to reduce the time and cost of deployment of the analytical solution.

On one hand, all of today's digital economy is becoming progressively dependent upon infrastructure such as hardware, software, telecoms, networks, etc.; and more and more business is conducted over computer-mediated networks and commerce is transacted over the internet – for example, transfer of goods or legal contracts. On the other hand, software solutions are also undergoing a major transformation with increased computational capacity and new ways to source more data that fuels machine learning (ML) and artificial intelligence (AI). These have essentially made SAP review their analytics portfolio and make it compelling for the digital economy.

In view of the aforesaid trends and developments, all the leading IT product companies had started to review their portfolio and focus on solutions for management of data as well as processing capabilities to generate insights. Some examples are shared next.

Microsoft Corporation acquired ProClarity Corp. in 2006 for adding advanced analysis and visualization technologies; and in 2008 it acquired DATAllegro Inc. for strengthening its MS-SQL Server with flexible software architecture and more recently in 2015 acquired Revolution Analytics R-based analytic solutions that can scale across large data warehouses and Hadoop systems and can integrate with enterprise systems. Similarly, *International Business Machines Corporation aka IBM* acquired Informix in 2001 and added the capability for high-performance domain-specific queries and efficient storage for datasets based on semi-structured data, time series, and spatial data. In 2004 they acquired Alphablox, Cognos in 2008, Netezza in 2010, and StoredIQ in 2012 to ensure their portfolio of solutions remained competitive and relevant for their clients.

In recognition of the opportunity and retaining its leadership in analytics solutions, SAP acquired BusinessObjects (BO) in 2007, the leading product then in the BI space. SAP further consolidated its position by acquiring Sybase in 2010 while introducing HANA, a new generation of an in-memory database with MPP capability. In 2013 SAP acquired KXEN, a leading product in predictive analytics with self-service features. The portfolio was further enriched by acquisition of RoamBi and Altiscale in 2016 to fill the whitespaces of mobile access to analytics and big data as a service offering.

The Evolution of SAP's Portfolio

With acquisitions, SAP's portfolio and offerings for analytic solutions had ballooned; and aside from causing confusion to clients and SI partners, its maintenance and sales were also challenged. SAP therefore started to rationalize its solutions and offerings. Following is an overview of the evolution of SAC within SAP's portfolio of solutions for BI requirements:

- Post-acquisition of BO, SAP engaged in integrating it within its portfolio and around 2010 started to present its solutions for specific BI tasks like Crystal Reports for reporting, Xcelsius for dashboards and visualization, and later – around 2012 – recommended its clients to start using SAP BO Design Studio instead of Bex Web Application Designer.

- With the introduction and successful adoption of HANA enabling advanced analytical applications, SAP started to redevelop all its solutions keeping HANA as the base for all its solutions. After acquiring Sybase and KXEN, around 2015, SAP recognized the need to converge its SAP BI client portfolio by announcing that SAP Lumira would become the mainstay for data discovery and analysis, while Design Studio would become the mainstay for all dashboard and analytical application development.

- In early 2015, SAP introduced SAP Cloud for Planning, and with its success, along with a rich road map, it evolved into SAP Cloud for Analytics and later rebranded to **SAP Analytics Cloud**.

Figure 1-3 depicts the evolution of SAC.

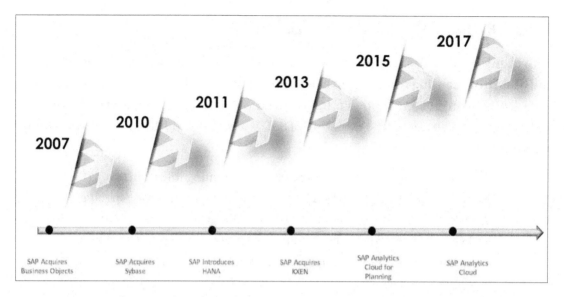

Figure 1-3. *Evolution of SAP Analytics Cloud*

SAP Analytics Cloud (SAC) is a single solution as a best-in-class Software-as-a-Service (SaaS) solution that combines all the analytics functionalities (planning, predictive, business intelligence) in one intuitive user interface, saving time and effort while making better decisions.

On the success of its ERP, SAP has been focusing on enabling enterprises to empower business users to make effective business decisions. In the digital economy, there is increased computational capacity with new ways to source more data. Managing, processing, or abundance of data is plausible by application of machine learning (ML) and artificial intelligence (AI); hence, alongside the transformation of its software, SAP was pushed to overhaul its offering for analytic challenges of the digital economy.

SAP has been at the forefront of the Enterprise Technologies and identified its offering for "Intelligent Enterprise" in three building blocks, namely "Intelligent Suite" consisting of the Digital Core (read S/4HANA), CRM, SRM, and suite of SAP products; the "Digital Platform" consisting of "Data Management" and "Cloud Platform"; and the "Intelligent Technologies" consisting of Machine Learning (ML), IoT, and Analytics. Business intelligence is a key component of the Intelligent Technologies.

Having learned about the positioning of analytics as one of the three pillars of SAP's offering for Intelligent Enterprise, let's explore capabilities of SAC that support the processes of an Intelligent Enterprise.

SAP developed SAC with the objective of offering a single solution for business intelligence and collaborative business planning. SAC delivers data discovery capability, be it on the data sources on premise or in the cloud, without moving, caching, or persisting any part or portion of the data into the cloud. The solution is built for the cloud and complemented with predictive analytics and machine learning technology. With SAC, SAP has unified the core domains of BI, planning, and is complemented by predictive analytics to deliver new capabilities such as simulation in BI, storytelling in planning, predictive forecasts in planning, or automated discovery in BI. The complementary capabilities from predictive and planning, on one hand, enable BI to shift from visualizing data to actually working with insights through ad hoc simulations, testing hypotheses, and planning for the future. On the other hand, it enables users to configure formulas for different accounts, manage currency conversion tables, as well as allocating values in planning.

Prior to SAC, SAP had multiple solutions in the portfolio (refer to Figure 1-4); and then around 2018, SAC became the primary solution combining capabilities of multiple solutions, namely SAP BO Explorer, SAP Roambi, SAP Lumira and SAP Analysis for OLAP.

Figure 1-4. *Rationalization of SAP BI portfolio*

Let's learn to appreciate a typical business situation in which a business user has to refer to multiple applications in order to achieve the desired analysis or planning, for instance, of emails, messenger services, ERP data, and business intelligence, among others. For example, in the planning process, the user may begin by reviewing historical and present data as a starting point possibly through an *analytical report*. The user then applies multiple formulas to generate forecasts and what-if scenarios, possibly done in a separate *predictive analytical* solution. After coming up with preliminary numbers, those are shared with management or other stakeholders for their review and approval; most often, sharing is done over a *collaboration tool or email or messenger*, and later they are presented or published to all the stakeholders. The situation described above showcases, how a user has to refer to multiple applications for sourcing and processing the information before it is published to the stakeholders, and that too while relying on the IT unit to support each of the tasks and processes before the output, can be meaningful for the relevant task or process.

Highlights of SAC Capabilities

In this section, we will learn about specific capabilities of SAC with underlying technical components and features. Figure 1-5 depicts capabilities of the SAP Analytics Cloud. We will briefly cover the All-in-One Platform, Augmented Analytics, Single Version of Truth, Anytime Available, Predictive Analytics, and Custom Analytics Application Design.

All-in-One Analytics Platform

With analytics forming a core pillar in SAP's Intelligent Enterprise framework, SAP's design view for SAC's architecture has been to promote a single platform based on HANA to encompass analytics, planning, and predictive analytics. With a view to eliminate referring to multiple applications, SAP brought together the capabilities of Analytical Solution and Planning in SAC as a single solution that enables performing each of the multiple tasks and process in the business in an effortless way, without being dependent on the IT unit. SAC is also the focus for SAP innovation and sees regular updates in terms of new technologies being rapidly incorporated into the platform. SAC is also native to the cloud, built from scratch on the SAP cloud platform that enables access to analytical functions from multiple devices. These features make SAC the tool of choice for bringing all analytics onto a single platform. Some of the benefits that SAC offers for bringing all analytics to a single platform are the following:

a. **HANA as the foundational platform:** SAC is built on the HANA Cloud platform. Having a versatile platform on the cloud with a proven database technology enables SAC to be equally flexible in terms of approaching analytics holistically. The HANA platform brings to the table robust data processing skills and new age storage technologies. Building on this framework, SAC delivers a complete packaged set of analytics functions that can be based out of a single data processing layer.

b. **Faster insight to action:** SAC's robust collaboration tools between end users and connectivity between tools ensure less time is spent on discussions and analysis of results. Decisions can be arrived at quickly with traceability within the system itself.

c. **Lower Cost of Ownership:** With the capacity to integrate multiple functions related to analytics onto a single platform, HANA provides the data processing capabilities, and the application layer of SAC brings together disparate functions. One of the primary benefits of a single platform is the total cost to the organization in terms of license and infrastructure. SAC's SaaS billing options ensure there is rapid scalability available whenever needed.

Augmented Analytics

SAC utilizes machine learning technology intrinsic to the underlying platform for delivering augmented analytics capabilities and uses techniques such as data mining, statistical modeling, and machine learning to present the end user with a forecasted value based on historical data. The augmented analytics capabilities of SAC are collectively known as "Smart Assist."

a. **Faster insights to action:** The augmented analytics of SAC enable end users to build stories using machine language technologies and converse with the analytics application in a natural language. The ease of exploring data brought about by Smart Discovery and Search to the Insight feature enables end users to focus on specific data points and explore data contextually with domain knowledge. Predictive Forecast and Smart Insights allow end users to rapidly convert analytics into actionable insights by creating a plan with the focus on forecasted values.

b. **Low learning curve:** SAC enables the use of modern advanced analytics and derive all the advantages without having to delve deeper into the understanding of machine learning and statistical models. Augmented analytics enable end users to include forecasts and simulations directly into their stories. Stories can be presented to management through the single SAC platform. With the low learning curve, end users can save time, avoid redundancy, and make the best use of available data to make data-driven decisions.

c. **Reduced redundancy:** SAC's Smart Discovery and Smart Insight features, when combined with the powerful storytelling capability, enable end users to build focused reports and presentations. Executive dashboards can be built directly over digital boardrooms, which allow for real-time data exploration. Deeper insights are presented by Smart Insights.

Single Version of Truth

SAC is built on the SAP HANA platform and is part of SAP Cloud solutions. SAP Cloud delivers multiple cloud-based solutions and services integrated into a single platform, allowing customers to integrate across multiple data sources including big data and streaming data, enabling a platform for Single Version of Truth. SAC's connections and modeling tools allow businesses to deploy a single tool for data exploration of all data sources. SAP sources also allow for live connectivity as well as seamless data flow across objects such as hierarchies.

a. **Low Data Footprint:**

The modeler in SAC enables data wrangling and blending. Heterogenous data can be rapidly transformed to provide end users with a strong semantic layer in the form of models. Models eliminate the need to store and process data across silos over multiple layers.

b. **Rapid Deployment:**

SAP has extended its best practices and standard business content to the SAC environment as well. SAP delivers business content across lines of business as well as across industry best practices. Out of the box, these content packages can be installed directly over the content delivery network and be rapidly deployed with minimal alterations.

c. **Quick Scaling:**

Enterprise Analytics Applications have traditionally been driven by licensing. Scaling up or down in terms of capacity is a major challenge for system owners to enable multiple layers of end users to be onboarded to these applications. SAC, being a cloud native application, allows rapid scaling in terms of capacity as well as users to the existing landscape.

Anytime Available

SAC is built on a cloud native solution, that is, built from the ground up on the SAP Cloud platform and maintained by SAP, ensuring around-the-clock access to data for analysis without downtimes.

a. **Native to Cloud**: SAC is a cloud native application built over the high-performance SAP HANA Cloud platform. The entire application can be accessed through a web browser without the need for installing a local application. Since there is no local installation, SAC can be accessed from a device with any OS or version. The browser also enables debugging of the application intrinsically without the need for an additional application. Also, the entire landscape, including regular updates, is maintained by SAP and the enterprises do not have to spend time and effort in upgrading.

b. **Mobile apps:** SAC provides a mobile app for both of the most popular mobile platforms, viz, iOS as well as Android. The iOS app has been available and has matured over the years whereas the Android app has been launched in Q1 2020.

c. **Content Network:** SAP provides pre-built out-of-the-box content for most Lines of Business (LOB)s and industries, which can be rapidly deployed through the Content Network. This enables end users to rapidly build and consume dashboards and stories from any location without the need to set up a separate development.

Predictive Analytics

SAP has integrated predictive capability within the SAC, allowing end users to create predictive scenarios and integrating those scenarios into analytical and planning stories. SAC delivers a custom-built predictive analytics capability solution with Smart Predict. Here are some of the benefits offered by Smart Predict:

a. **Informed Decisions:** SAC's Smart Predict technologies enable not only statistical analysis but also provide end users with an easy-to-use explainable technology tool, with a low learning curve. Enabling end users with Regression for predicting numerical values, Classification for binary decisions, and Time Series for forecast over time, informed decisions can help organizations plan for the future.

b. **Reduced Risk:** Smart Predict enables automated analysis of historical data with simple explainable technologies to enable end users to make informed decisions. Since the decisions are now driven by data and not on past experience, the probability of a disruption is very minimal, thus considerably reducing the risk involved.

c. **Improved Business outcomes**: With business plans targeted specifically for signal values based on forecasts, investments can be made specific to business lines and products, ensuring business outcomes. With specific predictive scenarios, organizations can drive better outcomes for business plans and improved investment results.

Custom Analytics Application Design

SAC enables building custom analytic applications. The custom analytic applications cater to specific requirements while relying on complex scripting to deliver a best-of-the-class experience to end users. Allowing customizations from data connectivity, discovery as well as user experience, while continuing to easily integrate with the available functionality like BI and planning, SAC delivers a complete package for end-to-end analytics.

a. **Flexibility:** SAC Analytic Applications ensure flexibility in terms of building applications where the standard reports and stories fail. With simple JavaScript-based scripting, the SAP Analytics Designer is able to fulfill most of the custom features to be built into the application. It also enables multiple components and widgets that can be placed onto the canvas and scripted as per requirements. Integrating into the existing architecture, the Analytic Applications display the utmost flexibility in terms of features.

b. **Reusability:** The components or widgets built within the canvas layout can be published as reusable components for use within the entire enterprise SAC landscape. SAC Analytics Designer also allows for the development of custom themes that can also be published for consumption. A considerable amount of time can be saved by reusing components. The SAC Analytics Designer also enables creating a new custom color palette that can be reused across other reports, stories, and applications. Custom CSS can also be included to enable further reusability within the SAC Application. This is especially useful when the organization has defined a set of colors for its brands as well as for the enterprise. Development effort can be reduced while increasing the efficiency of delivery of the applications to the end users by enabling reusability across the components and widgets.

c. **Insights to action:** SAC enables programmatic flexibility into the traditional information dashboard by embedding data actions into the dashboard. SAC thus enables Insights to Action within the landscape itself. This feature is especially useful for maintaining consistency and standards and enable a complete 360-degree execution for data actions. SAC's Analytics Designer offers a custom development environment that previously was available only with Lumira Designer and SAP Design Studio. Offering a complete custom development package, the Analytic Designer enables building applications to be built using connections to data sources or through file uploads. The Analytics Designer can be integrated with other components of the SAC like Planning, Stories, and Connections, as well as other web applications.

Further, SAC has an intuitive interface that enables a business user to perform tasks with ease, through functionality that looks and feels a lot like the tools they are accustomed to, thereby facilitating a faster adoption.

- – Spreadsheets without their typical challenges such as multiple copies or performance issues since the solution uses the SAP HANA Cloud Platform.

- – Mere selection of a range of cells and SAC proposes a most apt visualization for that data, along with multiple graphic options.

- – Sharing specific files, reports, or even cells of information with other users to discuss and create an action plan based on that discussion.

- – Managing a calendar of events that will help streamline the entire process.

- – Publishing presentations with the final information.

SAP's main focus has been in removing complexity and providing an all-in-one application, so it has therefore been continually enhancing SAC with many more such intuitive interface options that help companies make the most from their investment in SAC with the least time for adoption.

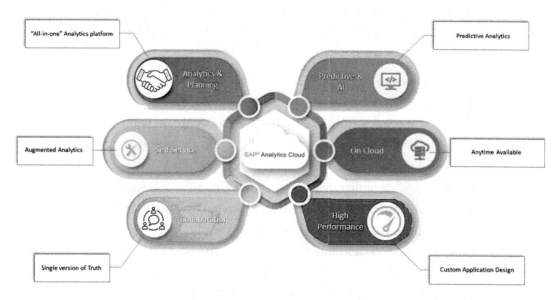

Figure 1-5. *SAP Analytics Cloud - Capabilities*

In summary, SAC is a Saas offering from SAP that combines the analytics and planning capabilities into a single solution. SAC consolidates data from all the applications and presents "smart analytics" for forecasting and what-if simulations through an intuitive and role-based interface. Analysts and planners in all lines of business are empowered with deep insights and data-based decision-making to act on issues collaboratively from anywhere.

Table 1-1 provides an overview of SAC capabilities mapped to underlying technical details.

Table 1-1. *Summary of SAC Capabilities and Its Technology*

Dimension	Capability	Technical details
	• Unified Platform for Business intelligence and Business planning processes. • Enable Data-Driven Decision-Making while simplifying planning and analysis. Embed Collaboration tools and analytics directly into planning processes to avoid switching between applications. • Enable Planning Capabilities with Predictive Forecasting and simulation including what-if scenario building for making decisions for future.	• Share stories and planning models across the enterprise. • Public and Private versions for planning models. • Value Driver Tree for Digital Boardroom: simulate and visualize important business drivers on large, triple interlinked touch screens.
	• Augmented Analytics to enable black boxed machine learning models for Guided Data Discovery.	• Smart Discovery to create machine learning-based stories with simulation. • Search to Insight to delivery Natural Language-based data discovery.

(*continued*)

Table 1-1. (*continued*)

Dimension	Capability	Technical details
	• On-the-cloud-based solution for scalability and flexibility.	• Independent planning application with live and import connectivity to multiple sources. • Hybrid planning with support for Embedded planning in S4HANA and SAP BPC.
	• Better business outcomes through smart decision-making. • Faster Actionable Insights.	• Run advanced calculations at blazing speeds with SAP HANA in-memory calculation engine. • Augmented Smart Analytics for enabling faster actionable insights.
	• Processes to boost collaboration and contextual decision-making Benefits. • Enable "Anytime Analytics" to allow stakeholders to access data across devices.	• Embedded collaboration within analytic processes. • Access to Analytics content and data exploration across multiple devices with mobile apps and web-based access.
	• Advanced analytics and scenario modeling without IT intervention. • Connect, prepare, and blend data from different sources. Create and enrich visualizations with insights from Big Data discovery. • Tools to draw intelligence from all data and present it in any application.	• Self-service storytelling capabilities. • Digital Boardroom for 360-degree view or organizational data. • Augmented Analytics to enable faster data discovery. • Smart Predict to create predictive scenarios for business forecasting.

Each of the above referenced capabilities and their technical aspects will be detailed in the following chapters of this book.

Summary

You have learned about the latest trends in analytics and how SAP is adapting to these trends. You further saw how SAC is SAP's solution for Analytics in the digital economy and forms a pillar of their Intelligent Enterprise offerings, concluding with a brief about SAC and its capabilities. In the following chapter, we will introduce you to requirements of a typical company operating in a digital economy and how SAC plays a vital in delivering the Intelligent Enterprise and empower decision-making for the stakeholders of that company.

CHAPTER 2

Business Scenario for Analytics Landscape Transformation

ABC Inc. is a multinational company with interests in multiple industries ranging from retail to real estate and manufacturing. Figure 2-1 represents ABC Inc. in a **graphical map.** In the recent past, it has acquired companies to consolidate its position in it's focus industry sectors and countries of interest. Leadership of ABC Inc. recognizes that to grow, remain competitive, and be successful across its business units, they need to remain agile and innovative. Therefore, ABC Inc. desires to empower each employee with information to be able to make quick and informed decisions for maximizing outcomes from their limited resources and be proactive in responding to opportunities in the marketspace.

ABC Inc. has been using SAP for their ERP requirements and multiple point solutions for analytical requirements. However, the acquired companies have had different solutions for ERP and analytics. ABC Inc. has already embarked on rationalization and consolidation of their IT landscape; for instance, implementation of SAP's latest version of the ERP solution on cloud across the entire organization and entities with the objective of standardizing and reducing the Total Cost of Ownership (TCO) on IT Infrastructure, Solutions, and Services.

© Vinayak Gole, Shreekant Shiralkar 2020
V. Gole and S. Shiralkar, *Empower Decision Makers with SAP Analytics Cloud*,
https://doi.org/10.1007/978-1-4842-6097-5_2

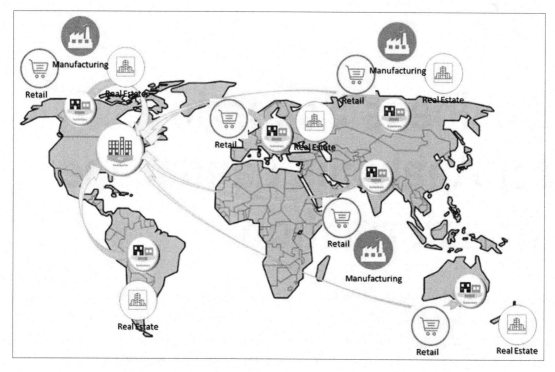

Figure 2-1. ABC Inc

Customer's Current Landscape and Pain Points

Currently ABC Inc. faces multiple challenges, mostly emanating from a variety of outdated tools, applications, and legacy solutions, be it in the software or hardware. Due to priority and attention to grow through acquisition, inefficiencies have cropped up, some due to deficient integration of the different systems, specifically from the acquired companies and the rest due from ever-increasing complexity in management of the entire IT landscape.

The flow of information is mired with a lack of synchronization throughout the IT infrastructure and remains a main cause for difficulty in managing the everyday business processes, leading to complications and negatively impacting the overall efficiency of the business.

Acquisitions of companies in similar lines of businesses, sometimes in the same geography, have added to factors that put spokes in the ability to have a single version of truth. For instance, there are significant duplicate records of customer information within ABC Inc. and an acquired company, resulting in a high redundancy of data scattered across various systems and applications in different formats, causing inconsistency in data and gaps in expected numbers and results. Further the complexity of the IT systems has become inflexible, lacking responsiveness to the dynamics of the ecosystem, for instance, changes in policies, procedures, and guidelines mandated for business, either by regulatory authorities in the markets or due to internal reasons. In addition, the variety of outdated tools, applications, and legacy solutions add delay in onboarding any business user, either new to the company or new to the role within the company.

Ever-increasing TCO for the portfolio of analytical applications poses a major issue apart from it lacking any of the modern capabilities and features that enable use of Advanced Analytics or mobility. Complexity of the systems also reduces agility in turning around any new requirements expected by the business user, which leads to erosion of reliance on Analytical Application by the business user community.

Listed below are some of the major challenges that ABC Inc. is dealing with.

Lacking Single Version of Truth

Rapid business growth of ABC Inc. as well as aggressive acquisitions have resulted in diverse information and systemic inability to integrate data sources, creating a lack of trust in corporate data. Questionable data quality, compromising reporting accuracy, forces a huge effort in reconciliation and so traceability is a major challenge. Reporting/analyzing across multiple systems or data sources is their biggest challenge. The problem is this data is spread across a variety of different systems and software – data stored in various ERP systems, CRMs, databases, and Excel spreadsheets. With data spread across multiple systems, getting the information is an arduous task. Decision support therefore consumes an enormous amount of attention and decision-makers often lack confidence in the results. Improved data quality, allowing users to trust the data for decision-making is paramount for ABC Inc. In present times, the wider business is running reports and pulling their own data and then ending up with multiple versions of the truth within the business; the other main challenge is around just the sheer time and productivity lost. It is normal to engage an analyst for reconciling and creating a consensual version of the report, anywhere between two and four hours, depending on how responsive the legacy ERP system is and data manipulation in Excel.

Lack of Standardization

Due to different solutions and technologies across ABC Inc., there's no standardization, be it calculation of KPI or definitions. Multiplicity of systems compounds the problem of data redundancy: scattered across silos and systems within the enterprise IT landscape. Lack of reusability results in redundant work efforts. Further the TCO is high due to management of multiple applications needing different vendors and suppliers, partners, and also the labyrinth of hardware and software solutions.

Lack of Collaboration

Business planning and decision-making lack collaboration are a cause for delays and high inefficiencies emanating from multiple applications. Sales staff at ABC Inc. spend a lot of time in the office analyzing data instead of focusing on their markets; so they have to then collect and collate data to prepare and validate the data over email with their sales managers, who do so in turn with their national managers, to get a quick summary dashboard for global management to get a sense of the business transacted on the previous day. Similarly, their finance and other analysts who need to actually get down into nitty-gritty details of the data look at the previous sales, what products were bought, and where they were being bought. They desire the solution to enable data-driven decision-making real, in a way that someone in Melbourne and somebody over in the United States in a meeting refers to and uses – the same tool, looking at the same data in real time.

Outdated System

One of the most occupying challenges is slow reporting performance leading to a number of internal challenges, severe reporting lags including fragmented processes, and poor interdepartmental communication. Enterprise information solutions are difficult to navigate and only accessible to those with advanced technical knowledge. The solution is not in league with analytical solutions used by the competition of ABC Inc.; for instance, the level of use of mobile devices is extremely low, and the use of machine learning or predictive algorithms are nonexistent. Advanced analytics is a black box to most of the IT team, so data scientists and experts are engaged for all such needs on an ad hoc basis adding to the increase of TCO and an inability to be agile and innovative. Due to the lack of augmented analytics, end users spend considerable time in data preparation and reconciliation. There are many additional challenges that ABC Inc. deals with; for instance, business planning is done in Microsoft Excel, and its iteration takes a huge toll on people involved, lacking collaboration. The solutions are not even meeting operational expectations, let alone enabling a competitive edge from the vast pool of information.

Lacking "Anytime-Availability"

Legacy applications are hosted on archaic and outdated infrastructure that need perpetual maintenance, causing a lack of availability on a regular basis. Further, due to complexity of the landscape, any change or new analytical requirement takes considerable time to respond to, and many end users therefore rely on using local solutions and working through their own on Microsoft Excel-based work. The current landscape is batch driven, and hence senior management lacks the availability of real-time data for critical business decisions. Overall, the system is extremely sluggish in response to meeting new requirements. Lack of access at any point in time over a secure landscape has seen ABC Inc. suffer in terms of raising the bar for timely decision-making across the organization.

Lacking "Mobility"

In a fast-paced, cutthroat digital economy, attributed to more mobile devices in the world than human beings, businesses must be able to make data-driven decisions empowered by insights in real time without limitations on mobility. Most of the tools and applications within ABC Inc. are not mobile compatible and necessitate the use of PCs, for example, dashboards access in the office and outside. Further the tools and solutions have very limited capability to customize apps. ABC Inc. executives are forced to produce work, planning, sharing information, or collaborating on projects in meeting rooms or using desktop PCs alone without access to data for critical decision-making.

Lacking Capabilities of Predictive Analytics

In an ultra-competitive world of business where identifying upselling and cross-selling opportunities is key for business development, ABC Inc. is faced with the crucial lack of predictive analytics features. Depriving end users of the power of data science capabilities has seen competition converting many business opportunities. Moreover, data at ABC Inc. is spread across multiple systems and landscapes without a clear data lineage. This siloed data bolstered by questionable quality has previously thwarted any attempt by ABC Inc. to introduce data science capabilities into the complex analytics landscape. Lack of insights from predictive analytics further hamper decision-making and planning capabilities at ABC Inc., thus resulting in increase of operational and marketing costs for the organization.

Lacking Rapidly Deployable Custom Analytics Applications

With a global footprint, ABC Inc. has multiple teams with diverse expectations from the analytics landscape. Some of the requirements demand a customized approach to analytics and can only be fulfilled by custom scripting. However, for enabling customized applications, data quality and single version of truth have been compromised with teams building their own version of analytical applications. The overburdened IT team is unable to cope with requests across multiple applications.

ABC Inc. is thus faced with the problem of having to deal with multiple customized applications over multiple sets of data, resulting in increasing operational costs and maintenance schedules.

Lack of Reusable Components

To fulfill custom application development needs, the teams at ABC Inc. have invested in building their own analytical applications. Due to the lack of collaboration, this has resulted in redundant components being developed, leading to a snowballing effect in the investments made by ABC Inc. in the development of analytical applications. Lack of reusability has taken a tremendous toll on the application maintenance team with significant efforts spent in rework and impact analysis for any change in the applications. This has led to ABC Inc. investing significantly in IT teams and systems without deriving true value from the investments. This has also discouraged adoption of self-service analytics within the organization with end users relying on custom applications instead of a single platform for all analytics.

Lack of Robust Security Architecture

ABC Inc. has primarily seen all its applications being available on premise with security deployed across the significant checkpoints. However, with the rapid rise in internet technologies, ABC Inc. has been witnessing a steady rise in cybersecurity issues. Without a robust security architecture, the analytics landscape always risks loss or unavailability of data for crucial decision-making capabilities. This is forever a risk that could prove to be detrimental in the ultra-competitive business environment in which ABC Inc. operates.

An overview of the challenges faced by ABC Inc. are represented in Figure 2-2.

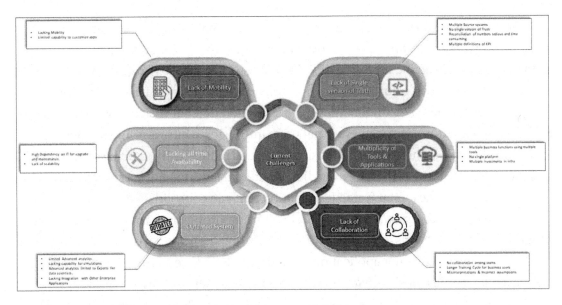

Figure 2-2. *Challenges of Analytic System at ABC Inc*

In context of the above-described scenario, ABC Inc. needs to transform their Analytics landscape and acquire the most appropriate Solution for dealing with challenges and capitalizing on opportunities.

Customer's Expectation from Analytics Landscape

A myriad of challenges faced by ABC Inc. from the existing Analytical Landscape has motivated them to identify a solution that more than meets their vision of having an application that ensures a *"Single Version of Truth"* with all-time access to business-related insights from one platform that will empower decision-makers to boost overall business performance while quashing the roadblock of poor data performance.

The solution should seamlessly integrate with their other Enterprise Applications, including ERP, and should *run on cloud* with optimal costs. The solution should be capable of enabling the organization to realize its aspirations of being an Intelligent Enterprise, that is, a modern data and analytics platform providing a competitive edge by providing the right information to the right people at the right time, along with multiple analytical capabilities coming together as part of the decision-making process and empowering the business user to deal with challenges and capitalize on each of the opportunities.

ABC Inc. desires to centralize their reports in a way that all reconciliation or consolidation gets done on the server overnight, and when they start the day each morning, the emails in the inbox have all the analysis they need to provide their executives – having that information at their fingertips at the start of every day. Further ABC Inc. management is keen that they have a solution that has the capability to consolidate data from diverse operational units into a single business intelligence reporting platform at the pace of the business instead of the business being chained by their systemic limitations, thereby providing greater insights to end-user teams across the world. ABC Inc. desperately needs a solution that can consolidate data from diverse operational units into a *single* business intelligence reporting platform to provide greater insights to end-user teams across the world.

ABC Inc. desires that the application should be able to access trusted data in real time, even from sales reps' *mobile* devices – improving decision-making, guiding product development, and helping customers by fueling customer demand and enabling business growth. This will lead to an increased ability and agility to help ABC Inc.'s customers.

The solution should allow users to connect all their data sources and sense beyond the numbers, discover new relationships, and *detect* trends to be able to take the guesswork out of each business decision. The solution should capitalize on the power of the emerging technology and enable an enterprise to respond rapidly to market needs and innovation and *anticipate* such needs by sensing important developments in internal and external environments in real time. The solution should also enable cutting-edge features like predictive analytics and augmented analytics to enable critical decision-making capabilities.

ABC Inc. aspires to adopt a *data-driven decision-making* and therefore deploy the right solution to benefit teams across an organization, allowing all users to easily collaborate and interact with, visualize, and communicate their data. A software solution that is capable of delivering insightful and actionable information without needing a data scientist to collect, prepare, and analyze complex data and process it into reports that management can understand.

It is recognized that with deployment of an analytics solution that ensures easy and fast-paced transformation to *self-service analytics* will not only reduce the TCO but eliminate the need for expensive IT support and empower business analysts with building their own reports and dashboards, along with easy drag-and-drop interfaces requiring little training and no prior data analysis or SQL skills, rather than waiting on IT to develop everything for them. This results in an agile, innovative, and cloud BI software solution to allow businesses to acquire the data they need without requiring anyone to wait and dole out money for deployment. Further, ABC Inc. is keen that their IT costs are part of the ongoing operational expenses by paying a subscription cost rather than one massive big capital expense requiring multiple stakeholders, including board members. While enabling self-service, the solution at the same time should also facilitate custom application development with reusable components, thus reducing redundancy drastically across the organization.

Reduced IT support for applications run on the cloud. lowers TCO for the company specifically for the business intelligence platform.

In view of growing data security-related challenges, ABC Inc. needs to ensure that data is secured from any potential business intelligence *security* issues by keeping that data safe and secure at all times.

Figure 2-3 summarizes the expectations of ABC Inc.

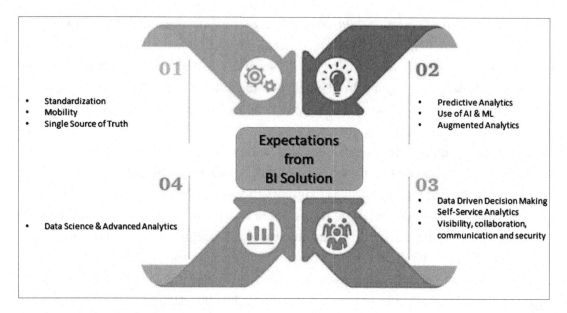

Figure 2-3. *Expectations from Analytic Solution*

Here's a quick reference (Table 2-1) on capabilities and features of SAC that are mapped to the challenges and opportunities of ABC Inc. in support of the evaluation and choice of SAC.

- Ability to see trends and spots gaps visually. Drastic reduction of manual effort in data collection, analyses, and alignment

- Reduced time spent on data validation and cleansing.

- A single version of the "truth" all contained in a single system to be used for reporting and operations planning.

- Allows for greater visibility of the forecast and fosters engagement across the Organization. Establishes monthly reviews prior to releasing the final version.

- One solution for data collection, manipulation, and reporting. Full visibility from browser or mobile device.

Table 2-1. *SAP Analytics Cloud Features Mapped to Challenges and Opportunities at ABC Inc*

Challenges and Opportunities at ABC Inc.	SAP Analytics Cloud Features
1. Multiple Source systems 2. No single version of Truth 3. Reconciliation of numbers tedious and time consuming	1. Connections 2. Models
1. Multiple business functions, multiple tools 2. No single platform, multiple investments in infra 3. No collaboration among teams 4. Incorrect numbers for Management	1. Stories 2. Planning 3. Digital Boardroom
1. Steep learning curve for business users 2. Incorrect assumptions and conceptions about analytics	1. Search to Insight 2. Smart Insights 3. Smart Predict 4. Smart Discovery
1. Deep Dependency on IT for upgrade and maintenance 2. Room for human error 3. Limited scaling 4. Limited handheld analytics	1. Maintenance by SAP 2. Unlimited scaling 3. Lower TCO 4. Mobile app
1. Predictive analytics for data scientists. 2. No room for citizen data scientists within the org	1. Easy-to-use Predictive analytics
1. Limited capability to customize apps 2. No single platform to integrate with enterprise analytics	1. Custom application development 2. Easy-to-use scripting
1. Global teams with disparate hierarchies 2. Limited integration with ERP	1. Users, roles, and teams 2. Integration with SAP security

As we described ABC Inc. as a representative of a typical enterprise and more specifically of the details of the challenges and opportunities, we have portrayed a real-world scenario that compels the choice of the SAP Analytics Cloud. In the following chapters, we will pick each of the challenges and opportunities and resolve them by learning about the configuration of capabilities and features of SAC through a step-by-step procedure to empower each stakeholder to make informed and timely decision-making while realizing their vision of an intelligent enterprise for ABC Inc.

Summary

In this chapter we learned about a typical Business Scenario for Analytics Landscape Transformation represented by ABC Inc., all the more so in its business context and rationale of an urgent need for a right-fit Analytics Platform for ABC Inc.

We then learned how features of SAC are mapped to challenges and opportunities at ABC Inc.

In the following chapter, we will learn about how SAC will serve the need of enabling "Single Version of Truth."

CHAPTER 3

SAC for Enabling "Single Version of Truth"

Rapid business growth of ABC Inc. coupled with them aggressively acquiring companies and lines of businesses across the globe has resulted in a huge diversity of information and systems. The problem is this data is spread across a variety of different systems and software: data stored in various ERP systems, CRMs, databases, and Excel spreadsheets. With data spread across multiple systems, getting information and drawing insights is an arduous task, which can lead to the inability to integrate data sources, culminating into a lack of trust in corporate data. Reporting/analyzing across multiple systems or data sources remains ABC Inc.'s biggest challenge. Questionable data quality and compromised reporting accuracy result in significant time and effort invested in reconciliation of information while traceability poses an equally major challenge. It is normal to engage an analyst for reconciling and creating a consensual version of the report, anywhere between two and four hours, depending on how responsive the legacy ERP system is and the data manipulation in Excel. Decision support therefore consumes disproportionate attention and decision-makers often lack confidence in the results. Improved data quality, allowing users to trust the data for decision-making, is paramount for ABC Inc. In present times, business units are running reports and pulling their own data, triggering multiple versions of the truth within the business; the other main challenge is around just the sheer time and productivity lost.

For the purpose of Enterprise reporting, when data is aggregated and moved across organizational hierarchies, organization leadership has been faced with multiple values for the same metric. Business KPIs instrumental in driving business direction and vision have been found to be faulty, requiring multiple iterations in arriving at an agreeable value. ABC Inc. has been faced with effort leakage in rework and delay in arriving at critical business decisions.

© Vinayak Gole, Shreekant Shiralkar 2020
V. Gole and S. Shiralkar, *Empower Decision Makers with SAP Analytics Cloud*,
https://doi.org/10.1007/978-1-4842-6097-5_3

Reconciliation of records and data is not only time consuming but also hinders corporate functions for planning and budgeting. For example, employee data for the organization resides in an independent human resources (HR) landscape. This landscape has its own processes and reporting capabilities that work flawlessly within the boundaries. However, if the CFO requests data analysis on how employee training costs have been affecting the organization's bottom line, employee data would need to be merged with financial data. Further drilling down and identifying the exact training courses that have the maximum impact vs. demand becomes a typical case for having a single platform, which can rapidly wrangle data and deliver the analysis on demand.

Note   ① Data wrangling, with respect to SAP Analytics Cloud (SAC), implies connecting or uploading data from different data sources, resolving quality issues, and enriching data or changing the layout of data. Data wrangling is an essential component of data preparation to provide a complete and clean dataset that can be used further for data analysis, trend detection, and critical decision-making within minimal time. Since the dataset is already prepared and ready for consumption, crucial time can be spent on analysis rather than data preparation.

Thus, ABC Inc. has been struggling with multiple internal and external systems to arrive at a single version and definition of truth. However, the underlying definitions and variances in these definitions have only resulted in chaos. Finance data reconciliation takes weeks, and interdepartmental collaboration is flaky and intermittent. This has resulted in a chaotic reporting mechanism that has left management unsure which numbers to trust for their decision-making process.

As we have already understood from the section "Customer's Expectations from Analytics Landscape" in Chapter 2, a primary requirement is to have a centralized reporting architecture. The Analytics Landscape should be able to meet the needs of a diverse user set including the executives being able to access the reports at any point of time. The platform should also have the capability to integrate data across multiple operating units as well as lines of business into a single analytics platform.

A modern analytics landscape, among other things, should be able to provide a mechanism to pull data from multiple systems to present users with a single version of truth. Connecting to these systems is only half the battle won. The analytics landscape should be able to pull each unique slice of data from disparate systems, cleansing the 'dirty' data, and publish information that enables decision-makers with a single clear version of truth: a truth that stays consistent across systems and across time and which can be reliably used to drive data-driven decisions.

One of the primary requirements from ABC Inc. has hence been the implementation of an analytical tool that will offer a "Single Version of Truth" across the organization. The tool should be capable of traversing multilayered data to arrive at acceptable values across the organizational KPIs delivering the ability to power critical business decisions. This expectation is detailed in Figure 3-1.

Figure 3-1. *ABC Inc. Requirements from the Analytics Landscape*

ABC Inc.'s technology landscape comprising multiple data sources should be able to converge onto a single reporting tool to deliver aggregated enterprise data sources for reporting and analysis. The tool should hence be able to provide connectivity to multiple data sources including cloud sources. The tool should be able to connect and display correctly organizational hierarchies as well as other elements from within the enterprise.

To summarize, ABC Inc. is looking for a single platform that can connect to multiple sources, wrangle datasets, and deliver content near real time across multiple devices, augmenting their business decisions.

SAP Analytics Cloud – Single Version of Truth

SAC provides a single cloud-based platform with native connectivity across sources to enable enterprises to build robust, well-defined KPIs. Connection options to multiple systems to pull data into a single repository allows SAC and publish it to the decision-makers.

SAC offers distinctive benefits to enterprises seeking to have a "Single Version of Truth," especially with SAP source systems. Some of the benefits for ABC Inc. are the following:

1. **Singular Platform:**

   SAC is built on the SAP HANA platform and is part of SAP Cloud solutions. SAP Cloud delivers multiple cloud-based solutions and services integrated into a single platform allowing customers to integrate across multiple data sources including Big Data and streaming data. Offered as Software as a Service (SaaS), built natively on SAP Cloud Platform, SAC allows organizations to close the gap between transactions, data preparation, analysis, and action, providing all analytics capabilities in one offering.

   SAC's incumbent connections and modeling tools allow businesses to deploy a single tool for data exploration for all data sources. SAP sources also allow for live connectivity as well as seamless data flow across objects such as hierarchies. Security objects are also tightly integrated with SAP source systems. The import data feature provides additional flexibility to end users to integrate data across sources.

   With all the above features, SAC provides a single platform for integrating data across sources and enabling businesses to deliver a single version of truth across the enterprise.

2. **Low Data Footprint:**

   The modeler in SAC enables data wrangling and blending. Heterogenous data can be rapidly transformed to provide end users with a strong semantic layer in the form of models. Models eliminate the need to store and process data across silos over multiple layers.

HANA's inherent data management capabilities, which forms the underlying database for SAC, bring in best-in-class data compression and storage to enable enterprises to work with lean data transformation.

3. **Rapid Deployment:**

   SAP has extended its best practices and standard business content models to the SAC environment as well. SAP delivers business content across lines of business as well as across industry best practices. These out-of-the-box content packages can be installed directly over the content delivery network and rapidly deployed with minimal alterations.

   Additionally, SAC being a cloud native application completely eliminates the need for a requisition, install, and deploying an application.

4. **Quick Scaling:**

   Enterprise Analytics Applications have traditionally been driven by licensing Scaling up or down in terms of capacity and is a major challenge for system owners to enable multiple layers of end users to be onboarded to these applications. SAC, being a cloud native application, allows rapid scaling in terms of capacity as well as user addition. As more business functions bring in disparate datasets, SAC enables its augmented analytics capabilities to assist end users delve deeper into data and deliver reliable KPIs.

We have learned about the benefits that SAC will offer ABC Inc. in their journey toward building a Single Version of Truth. Let us now learn about the building blocks that enable SAC to deliver the key requirements of a single version of truth, which are the following:

- Connections
- Models

The process for integrating the building blocks of connections and models for creating a single version of truth for data across multiple sources is shown in Figure 3-2. The process to create a single version of truth consists of the following components:

1. Create

2. Connect

3. Prepare

4. Model

5. Share

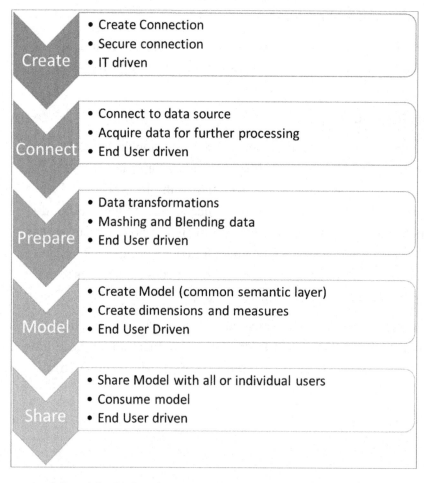

Figure 3-2. Building blocks for data integration and creating single version of truth

Note   Since SAC is a constantly evolving product with SAP introducing new features every quarter, some of the actual screens might be different from what are shown in this book.

Connections

Connections are pipelines that enable SAC to connect to data sources. The primary challenge of collecting data across sources is to navigate to the exact structure in the source that stores data, which could be a table, view, cube, etc. SAC enables this by providing native connectivity to multiple sources. This eliminates the need to install and set up compatible connectivity drivers. However, the administrator has to set up the connection manually by either entering the Service URL or Key/Secret/User/Password/ API Key/Upload file, depending upon the source system. This ease of connectivity enables SAC to connect, fetch, and wrangle data from disparate sources irrespective of how they store data.

As SAP continues to deliver better connectivity, as of March 2020, SAC provides connectivity to the sources shown in Figure 3-3.

Live	Import
SAP Cloud Data Sources	**Cloud Data Sources**
SAP Cloud Platform	SAP Cloud Platform
SAP Marketing Cloud	SAP Marketing Cloud
SAP S/4HANA Cloud	SAP S/4HANA Cloud
SAP Commissions	SAP Concur Solutions
	SAP Fieldglass Solutions
	SAP Sales/Service Cloud
	SAP Business ByDesign
	SAP SuccessFactors Solutions
	SAP SuccessFactors Workforce Analytics
	Dow Jones DNA
	Google BigQuery
	Google Drive
	Google Sheets
	Odata
	SalesForce
SAP On Prem Data Souces	**SAP On Prem Data Source**
SAP HANA	SAP HANA
SAP BW	SAP BW
SAP BW/4HANA	SAP BW/4HANA
SAP S/4HANA	SAP S/4HANA
SAP BPC Embedded	SAP BPC
SAP BusinessObjects Universe	SAP BusinessObjects Universe
	SAP Integrated Business Planning
	SAP ERP
Access via SDI	**Other Data Sources**
IBM DB2	IBM
MySQL	MySQL
Netezza	Netezza
Odata	Odata
Apache Hive	Oracle
AWS	CSV
SAP Vora	Excel
SAP Sybase ASE,IQ,ESP	Progress OpenEdge
Microsoft SQL Server	Microsoft SQL Server
Spark SQL	
Oracle	
Teradata	
MaxDB	
Facebook	
Google+	
Twitter	

Figure 3-3. *SAC connections provided by SAP*

Additionally, SAP partners also provide connectivity options. These options are shown in Figure 3-4.

Live	Import	
APOS	**APOS**	
DB2	Web Intelligence	
Denodo		
Microsoft Analysis Services	**CDATA**	
Microsoft Azure	Quickbooks	
Microsoft SQL Server	Microsoft SharePoint	
Mongo DB	NetSuite CRM & ERP	
Mongo DB Atlas	Oracle Marketing Cloud	
MySQL	Mongo DB	
Netezza		
Oracle	**DataDirect Cloud**	
Oracle Essbase	Eloqua	
Oracle Exadata	Google Analytics	
SAP IQ	Microsoft Dynamics CRM	
SAP SQL Anywhere	Hubspot	
Redshift	Marketo	
Teradata	SugarCRM	
Google BigQuery		
Cloudera	**EPI-USE Labs**	
Hortonworks	SAP Payroll	
Amazon S3	SAP Payroll config	
	SAP on-prem reporting	

Figure 3-4. *SAC Connections provided by Partners*

SAC Connections are primarily of two types. We will discuss them next.

Live Connections

A live connection enables SAC to connect to data sources and only import metadata instead of the entire dataset. Live data connections are near real time and provide up-to-date updates from the source systems. This type of connection uses the browser as a medium to connect to the source and retrieve metadata. The following are prerequisites that need to be set up on the source for setting up a live data connection:

1. Install and enable SSL certificate on the source system.

2. Set up the INA service from within the SAP source system.

3. Enable CORS (Cross Origin Resource Sharing) on the source.

4. Whitelist the URL for the SAC tenant.

One of the primary benefits of a live connection is it allows data to keep residing in the source, assuaging concerns about security. Also, since the data is processed at the source, the resources available in SAC are used optimally, enabling faster response times. However, live data connection models allow minimal transformations, which we will learn about in a step-by-step process in the upcoming sections.

Import Connections

An import connection imports data into the underlying HANA Database of SAC. On connecting to the source, a secure pipeline is established by implementing the SAP Cloud Connector, which acts as the gateway to the data. SAP Cloud connector can be downloaded from `https://tools.hana.ondemand.com/#cloud`

Import data connections and models enable much higher flexibility of churning data, enabling better analysis – however, at the cost of storing and processing data in SAC. We shall learn about import connections further in the next chapter.

Live Connection vs. Import Connection

We have explained the two different types of connections available in SAC in the previous subsection. In this subsection, let us learn about the basic differences between the two connections across the below parameters as shown in Figure 3-5.

Sno	Parameter	Live connection	Import connection
1	Speed	Slower	Faster
2	Type of Data	Metadata	Actual Data
3	Volume	Unlimited	Row limitations
4	Transformation	Limited	Multiple transormations available
5	Security	Data resides on customer Landscape	Data imported to SAP Landscape

Figure 3-5. *Live connection vs. Import connection*

- Speed:

  Since a Live connection does not import data, it has to rely on the data source for all transformations and data aggregations. Multiple factors like network latency, firewalls, and other security mechanisms contribute to data being delivered relatively slower to an Import connection. However, the difference in speed is minimal unless a large volume of data is used for modeling using a live connection.

- Type of Data:

  A Live connection extracts only metadata for building the model. The source data continues to reside in the respective source. For an Import connection, the entire data is imported to the underlying HANA database of SAC.

- Volume:

  Live connection can work with virtually unlimited data since the data is not imported. However, for an Import connection, since data is imported into the HANA database, it is limited. Some of the limitations for the most popular datasets are:

  - Excel and CSV: 2,000,000,000 rows; 100 columns

  - SAP BW, SAP Universe, SAP HANA: 100,000,000 cells; 100 columns

  - Other data sources: 800,000 rows; 100 columns

- Flexibility:

  Since a Live connection does not import data, it cannot be transformed or changed in a model. All transformations have to be done at the source. Import connection allows data to be transformed within the model. An Import connection is thus more flexible than a Live connection.

- Security:

  A Live connection does not extract data out of the source whereas
  an Import connection extracts data out of the source and brings
  it to the SAC servers. This is loaded into the underlying HANA
  Database. While using a Live connection, since data continues to
  reside at the source, all the underlying security continues to apply.
  A Live connection is thus considered to be more secure than an
  Import connection.

  Selection of the type of connectivity depends on the factors
  defined above. Customers, including ABC Inc., would need
  to evaluate their security policies as well as transformation
  requirements before finalizing the connectivity type.

Models

Models are the core building blocks of data analysis in SAC. They also enact the role
of a semantic layer enabling end users to explore data with ease even while having
minimal knowledge of underlying data objects. Models can be enriched with master
data, interspersed with data from multiple sources, and also enabled with data quality to
ensure clean data is always available for end-user analysis.

In SAC, models are of three types:

- Model Type - Analytic models:

  These model types allow end users to connect to data sources
  and cleanse and transform data to have a foundation for the story
  mode. These models allow primarily data exploration and self-
  service across the dataset they enable. They can be set up using
  both live and import connections.

- Model Type - Planning models:

  Planning models provide all the features of the analytic models,
  additionally allowing setting up of budgets and creating plans and
  forecasts. These highly specialized models allow independent
  planning as well as tight integration with back-end planning

systems like SAP BPC and SAP S4/HANA. These models further allow users to build comparative analytics combining planning data with enterprise data.

- Model Type - Predictive models:

  Predictive models are part of the Smart Predict capabilities of SAC and enable creation of predictive models over historical data. These models can be trained and set up quickly to provide enterprises to work on predictive capabilities.

In models based on import connections, data can be refreshed periodically. This process can be automated and scheduled as jobs. Jobs can be set up within models and can be monitored from the connections on which they have been set up.

Let us now learn about the core building blocks of models, which are the following:

- Dimensions

- Measures

- Hierarchies

- Transformations

Dimensions

Dimensions represent information about qualitative data and cannot be part of mathematical operations: for example, customer name.

The below are the types of dimensions available in SAC.

- Account:

  The Account Dimension is the primary dimension in a model. It can be considered as a grouping of all measures included in the model. All measures appear as members in a single column under Account Dimension. For example, if there are two measures – Sales and Revenue – they will appear as two separate members under Account Dimension.

  Unlike other dimensions, a model can have only one Account Dimension.

- Generic:

  A Generic Dimension is an object that represents qualitative data: for example, master data that does not change frequently.

- Date:

  A Date dimension holds dimensions of the type Date. These types of dimensions are especially useful while doing trend analysis and data exploration over a period of time.

- Organization:

  This is a typical type of dimension that maps to the organizational units from the source. Regions and operating units are typical examples of Organization Dimension.

- Geo Dimensions:

  By enriching location data with geo dimensions, map-based visualizations can be built along with features of further data exploration. Geo enriched data can also be used as location hierarchies for drilling through.

- Category:

  A Category Dimension, used in planning, is used to depict the data categories used in planning: for example: actual, forecast, and budget. Each change in the categories is stored as versions, allowing planners to compare data across versions and categories.

Measures

Measures store numerical values on which mathematical functions can be applied: for example, revenue. Measures in SAC are stored under the Account object type. An account is a holistic object that holds multiple measures under its umbrella.

Hierarchies

Hierarchies allow data drill down and enable exploration from top down or bottom up. Hierarchies can be built manually in a model or can be replicated from the source. A typical example hierarchy for HR would look like the one below:

- CEO

    - Director

        - Senior Manager

            - Manager

                - Team lead

                    - Team member

This type of a structure enables data exploration from the top to bottom of the entire organization. Hierarchy is a standard feature for all dimensions. SAC provides two types of hierarchies, viz, Level-based hierarchy and Parent child hierarchy. Let us learn about each of the hierarchy types in detail.

- Level-based Hierarchy:

    Level-based hierarchies are based on levels. The members within the hierarchy are of different types and are grouped across levels based on their similarity. Let us consider a Time-based hierarchy that has Year, Quarter, Month, and Week. Though there might be different years, they would all be grouped under year. This is shown in Figure 3-6.

Figure 3-6. *Level-Based Hierarchy*

A Level-based hierarchy helps end users explore data by drilling through the different levels and developing trends across levels.

- Parent Child Hierarchy:

  Parent child hierarchies are based on structures where all members are of the same type. For example, consider the organizational hierarchy. Even though everyone is of the same type, which is employee, the organization maintains a hierarchy across multiple layers. For example, see the employee hierarchy shown in Figure 3-7.

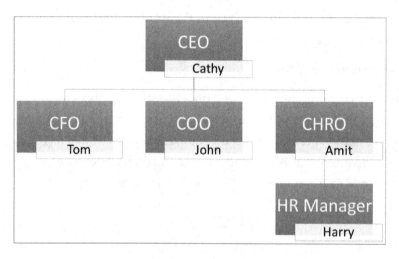

Figure 3-7. *Parent Child Hierarchy*

As can be seen in the employee hierarchy in Figure 3-7, the top layer is represented by the CEO, which is Cathy. The Parent CEO layer is followed by the child CFO, COO, and CHRO layers and so on.

Transformations

Data acquired into a model can be transformed using formulas and quantitative functions into a single version of truth.

Creating and Editing a Connection

SAC is a web-based application natively built on the SAP Cloud platform. Logging in to the application is enabled by a URL, a username, and a password, which is enabled when the end user requests access to the system. The URL also depicts where the SAC tenant for the customer is located in the SAP Cloud. We will learn in depth about how to access the SAC landscape through a step-by-step process in Chapter 6's section, "Enabling Anytime Analytics with SAC." We will also learn how to access SAC through the web as well as the mobile app delivered over iOS as well as Android devices.

In this chapter, we will focus on creating a connection to bring in data into a model and how to enable a Single Version of Truth for all analytics across the enterprise.

In this section, we will also learn to create a connection through a step-by-step process. Subsequently we will build a model and understand the nuances of creating one.

First, let us learn about the connections screen. To access connections, use the ❶ "**Connection**" option of the menu shown in Figure 3-8.

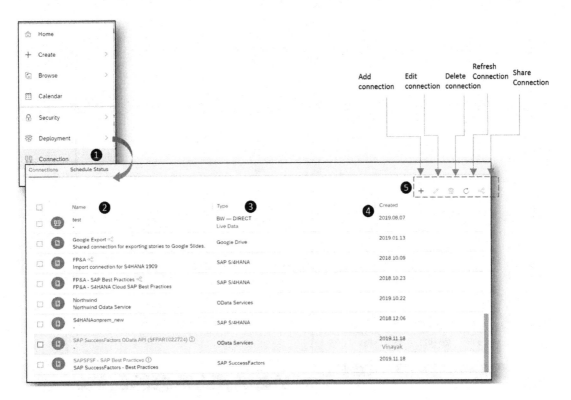

Figure 3-8. *Connection Screen*

The screen with all the connection options then opens. The connection screen shows the list of default connections. This screen shows the name of the connection as shown in ❷, type of the connection as shown in ❸, and creation information as shown in ❹. Other options available are as shown in ❺ to do the following:

- Add connection

- Edit an existing connection

- Delete an existing connection

- Refresh connection

- Share an existing connection

Create a Connection

Let us now learn how to create a connection. The steps to create a live connection are shown in Figure 3-9.

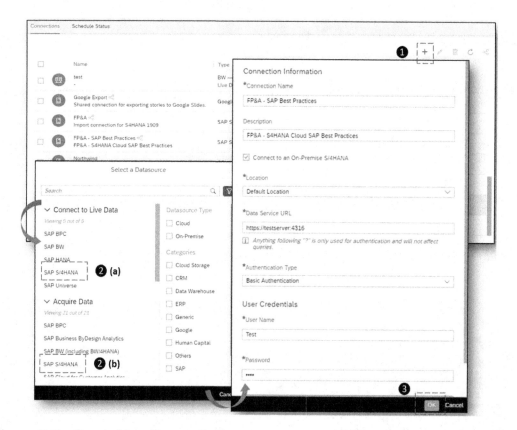

Figure 3-9.   *Create New Connection*

Step 1: From the menu, click on the '+' symbol as shown in ❶. The screen for creating a new connection comes up in a pop-up window. For creating a Live connection, follow Step a; and for creating an Import connection, follow Step b from the list below.

a.   Live Connection:

Select a data source as shown in Figure 3-9 under ❷(a) from "**Connect to Live Data.**"

b.   Import Connection:

Select a data source as shown in Figure 3-9 under ❷(b) from "**Acquire Data**" to bring up the dialog box for creating an Import connection, if the requirement is for an Imported data source.

Step 2: The next screen comes up as shown in Figure 3-9. Provide a name for the connection and fill in the rest of the information as expected in the subsequent dialog boxes. Select the username and password to be used for connecting to the source system. In case a cell is used, select the option from the authentication type dialog box. Click **"ok"** to create the connection. This connection will now be visible in the connections screen.

Edit a Connection

Many times, certain parameters need to be updated within the connection settings. These changes can be brought about by editing the connection. The step-by-step process to edit a connection is shown in Figure 3-10.

Figure 3-10. *Edit Connection*

Step 1: Select the connection to be edited. This is shown in Figure 3-10.

Step 2: Click on the 'pencil' symbol at the top right-hand corner of the menu.

Step 3: This will bring up the edit connection dialog box that will have all the information that has already been provided while creating a new connection. This information can be edited and changed as per the new requirement. Click **OK** to finish editing the connection.

We have learned in this section how to create and edit a connection. In the next section, let us learn about Jobs.

Jobs

Jobs enable automatic update of data at regular intervals into a model. Jobs enable loading of data as well as exporting data from a model as a CSV file. Since jobs enable brining data into or out of the SAC Landscape, they are available only with an Import Connection. Based on whether the data is being imported or exported, Jobs are of two types:

- Import Job:

  An import job enables bringing data into the model. It can be one time or can be scheduled to run at regular intervals. The job will use the underlying connection to connect to the data source and load data as per the schedule specified.

- Export Job:

  An Export Job enables moving the data from the model into a csv file or back into the data source. This is especially useful for planning models that have been integrated with other planning applications like BPC in the back end. Once the end user updates the plan, it can be sent back to the source through an export job.

Created jobs can also be monitored from the connection screen. Click on the **"Schedule Status"** tab ❶ in Figure 3-11, beside the Connections tab to check the status of the scheduled Jobs.

The window shows all attributes of the job including the connection, type, and which fields are being imported or exported. Select a job and click on the Bin icon ❷ to delete it and on the "**Refresh Now**" button ❸ to refresh the job manually. The Schedule Status for jobs is shown in Figure 3-11.

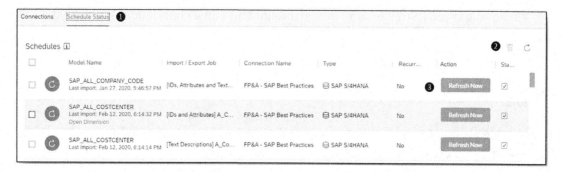

Figure 3-11. *Job Schedule Status*

And thus, connections enable us to connect to multiple types of data sources. In the next section, we will learn how to create a model and combine data from multiple data sources using connections to enable a "Single Version of Truth."

Create an Analytical Model

As discussed in the previous section, Models in SAC are one of the building blocks of data transformation. Encapsulating business logic in easy-to-understand wrappers, Models allow data across multiple sources to be combined, wrangled, transformed, and shared across the enterprise.

Since in this chapter we are focusing on the capability of a single version of truth, we shall discuss how to create analytic models only. Planning and predictive models will be discussed in subsequent chapters.

From the main menu, select **CREATE** and **MODEL** as shown in Figure 3-12 to start the process of creating a model.

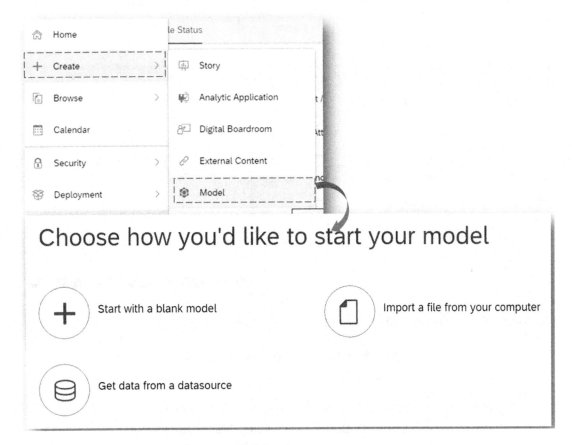

Figure 3-12. *Create Analytic Model Options*

As shown in Figure 3-12, this provides us with three options to create a model:

- Start with a blank model.

- Import a file from your computer.

- Get data from a datasource.

Let us learn about each option in detail to understand how models can be created.

Start with a Blank Model

This option creates a shell for creating a model. This model can then be enhanced by building dimensions and measures. The step-by-step process to create a blank model is shown in Figure 3-13.

Figure 3-13. *Start with Blank Model*

Step 1: Click on "**Start with a blank model**" as shown. This will bring up the model's list window as shown in Figure 3-13.

Step 2: Provide a name for the blank model. In this case we have given the name "**B_Blank_test**". Click **"OK"** to create the model. Since this is a blank model, it will not have any standard dimensions or measures. The end user can either add an existing dimension or can create a new one.

Step 3: Create a dimension by clicking on **"Add Dimension."**

Step 4: Select the appropriate measure and click **"Save"** to save the model.

Step 5: The next step is to load data into this model. Data can be loaded by connecting to the source and fetching the records via jobs.

Step 6: For this purpose, go to the **"Data Management"** tab as shown and create a job.

We have learned about jobs in the previous section titled **"Jobs."**

Import a File from the Computer

SAC allows models to be created by importing a data file from the local computer or from a file location. This increased flexibility is useful while enriching transactional data or for building a model that is primarily based on local data.

Figure 3-14. *Import file from Computer*

Figure 3-14 shows the process to create a model from an imported Excel file. Let us learn about the process step by step:

Step 1: In the model builder window, choose the option "**Import a file from your computer.**" This is shown in Figure 3-14.

Step 2: Click on the option to "**select a source file.**"

Step 3: This will open the **"File Explorer"** in the local computer. Select the file you want to upload to SAC for creating the model. In this case, we have selected the **"sales_demo_sample"** Excel file. Click **"Open."** While importing from a file, Excel (xlsx only) and csv files are supported.

While importing from an Excel file that has multiple tabs, select the sheet that has to be imported. Note that all tabs will not be imported to the model. While importing from a csv file, the delimiter has to be specified, which would differentiate between columns to be imported. Selecting **"Auto Select"** will enable SAC to select the best option for delimiting and loading the data as columns in the model.

Step 4: The next window comes up, which shows that the file has been successfully selected. Click "**Import**" to import the file into the SAC platform.

Note: Ensure that the source file is not open before importing into SAC.

Step 5: A message comes up successfully importing the file, which points out that all data might not have been imported into the system. Click **"OK."**

If the data in the file being loaded has a large quantity of data, SAC loads only minimal rows of data, which would serve as a base for further data transformations in the model. Business rules and transformations applied on a large quantity of data are not only time consuming but could also reduce efficiency for the end user building the model. SAC hence loads a part of the data that can be used for building the transformations and once all rules have been applied, during the final model creation, all data is uploaded, processed, and stored as columns in the model, which is shown in Step 6.

Step 6: The next window brings up the data that has been imported into the system, which can be wrangled and transformed. On completion, click "**Create Model**" to successfully create the model.

Get Data from a Data Source

SAC also allows creating a model by either using an import connection or by using a live connection to a data source. While building such a model, the same data source choices are presented as while building a connection. Likewise, a model of this type can be built on top of a live data source connection or an import connection. Let us learn about the steps to create a model from a data source.

Create Model from a Live Data Source

As we have already discussed in the section **"Connections"** and subsection **"Live Connections,"** a live connection brings only metadata into the model. Similarly, a live data source brings in only the metadata. Refer to Figure 3-15 to understand steps involved in creation of a model.

Figure 3-15. *Create Model from Live Data Source*

Step 1: Click on the "**Get Data from a datasource**" option.

Step 2: The connection screen comes up. Select the option to "**Connect to live data**" as shown in Figure 3-15.

Step 3: The connection window comes up. Select the type of the system to connect to. The system types can be one of the following:

- SAP BW for connecting to SAP BW-based systems including BW/4HANA and S4/HANA.

- SAP HANA for connecting to native HANA-based calculation views.

In this example we have selected SAP BW.

Step 4: From the connection list, select the connection. Note that only live connections will be available. On selecting the connection, select the data source. In this case we have selected a CDS view that has been exposed as OData for creating a live model. Additionally, other data sources can also be used like HANA view and Bex Query. For exposing a CDS view to be consumed, the annotation has to be added, which is "*@OData.publish: true*".

Step 5: The live model comes up in the modeling window. Click on "**Create Model**" to finish creating the model.

Create Model from an Import Data Source

An import connection imports data into the underlying HANA database in SAC. A Model can be created from an Import data source from the following:

- A new query

- Copying a query from an existing model

Create Model from a New Query

This option allows the user to create a new model by creating a query on the source while using an existing connection to connect to the source. Let us learn about the steps involved for creating a model from a new query. This is shown in Figure 3-16.

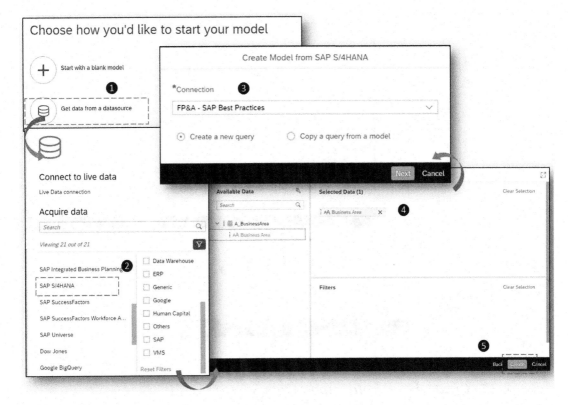

Figure 3-16. *Create model from Import Data Source - with New Query*

Step 1: Click on the "**Get data from a datasource.**"

Step 2: Select a source from the list of data sources listed under "**Acquire Data.**" In this particular case we have selected SAP S/4HANA.

Step 3: The connections window comes up which lists import connections. Select the data source. From the options below, select the options to "**Create a new query.**" Click **"Next."**

Step 4: The query window comes up as shown in Figure 3-16. Select the dimensions needed to complete the query click on **"Create"** to create the model.

Step 5: Click on **"Create"** to create the model.

Copy a Query from a Model

Let's see the final option for getting data (see Figure 3-17).

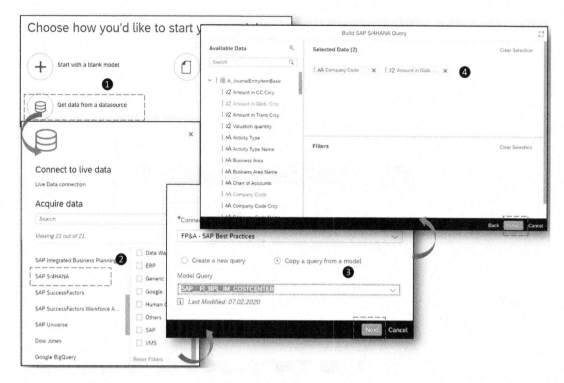

Figure 3-17. *Create model from Import Data Source*

Step 1: Click on the "**Get data from a datasource**" as shown in Figure 3-17.

Step 2: Select a source from the list of data sources listed under "**Acquire Data.**" In this particular case we have selected SAP S/4HANA.

Step 3: The connections window comes up, which lists import connections. Select the data source and select the option "**Copy query from another model.**" The list of models comes up. Select the model as shown in Figure 3-17. Click **"Next."**

Step 4: The query window comes up. However, this contains only those dimensions that have been selected in the underlying model. Select the required dimensions and click "**Create**" to finish the model.

* Note: Once a model is created, it cannot be edited or changed.

Models introduce a semantic layer in the data analysis process. A model can be reused across multiple stories, can wrangle data, and even provide data security. End users and developers who work with stories are always insulated from the nuances of data wrangling and blending and can consider a model as the base for story creation. However, a model need not be a prerequisite for creating a story. A story can be created by directly connecting to source data. In particular scenarios that demand analyses of data without the need for reusability, stories can be built directly on the data. This saves considerable time by reducing the effort spent on data preparation and model creation. Rapid insights for action can be achieved by delivering stories that are built on a dataset restricted to a use case and without the need for reusability. We will learn about some stories in detail in Chapter 4.

SAC Data Modeler View

As discussed in the previous section, Data Modeling is the process of enhancing available data by cleansing, transforming, and wrangling it to build a dataset that can be used in self-service analytics, stories, and dashboards. The Modeler View in SAC enables users to build models. The Modeler View enables further capabilities including setting of units and currencies. It also enables data transformation as well as resolving data quality issues. Custom calculated columns can be created for data that is not readily available, and the entire dataset can be combined with another dataset enabling data blending. We will learn about all of these features in this section as well as the menus available for the model creation. We have already seen how to build models. Let us now learn about the Modeler View and its components.

Top Menu Components

The Modeler in SAC is shown in Figure 3-18. Let us learn about the Top Menu components before moving to the side menu in the next section. Each of these components is essential for data wrangling and preparation to ensure the resulting model is ready for story creation.

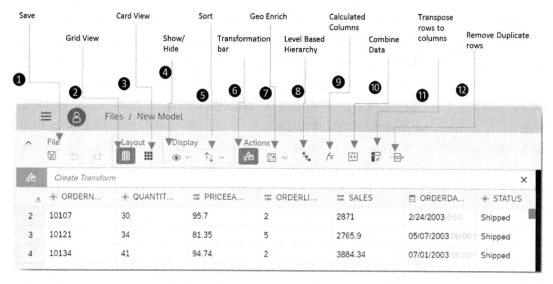

Figure 3-18. *Modeler Window*

1. **Save:**

The file menu has the option for saving the created model. The model is saved in the SAC platform. Any transformations done to the data are saved as well.

2. **Grid View:**

To view the data in the model, SAC offers two views:

- Grid View

- Card View

Grid view shows the data in the form of a standard table. This view is shown in Figure 3-18. In this view, the transform window is also available to create data transformations.

3. **Card View:**

The Card view shows columns as cards. This view brings up data quality of and provides statistics on how many rows fulfill the criteria for clean data. The Grid view is shown in Figure 3-19.

Figure 3-19. *Modeler Card View*

The green bar indicates that there are no data quality issues for the specific columns.

4. **Display Columns:**

This option allows the user to select which columns to be made visible in the model. Only the columns selected are visible, and the rest would be hidden as shown in Figure 3-20.

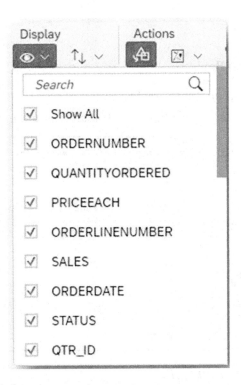

Figure 3-20. *Display Columns*

5. **Sort:**

This option allows sorting of a particular column in ascending or descending order. This is shown in Figure 3-21.

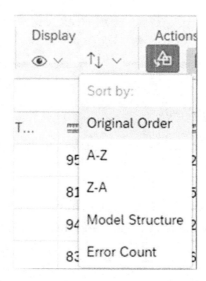

Figure 3-21. *Sort Order*

6. **Transformation Bar:**

The transformation bar enables transformation of data within the model. The options shown in Figure 3-22 are available for transforming the data.

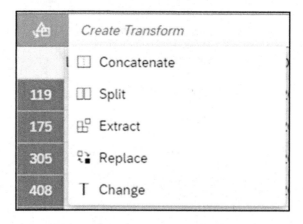

Figure 3-22. *Transformation Options*

In Figure 3-23 we learn about each option for transformation in detail.

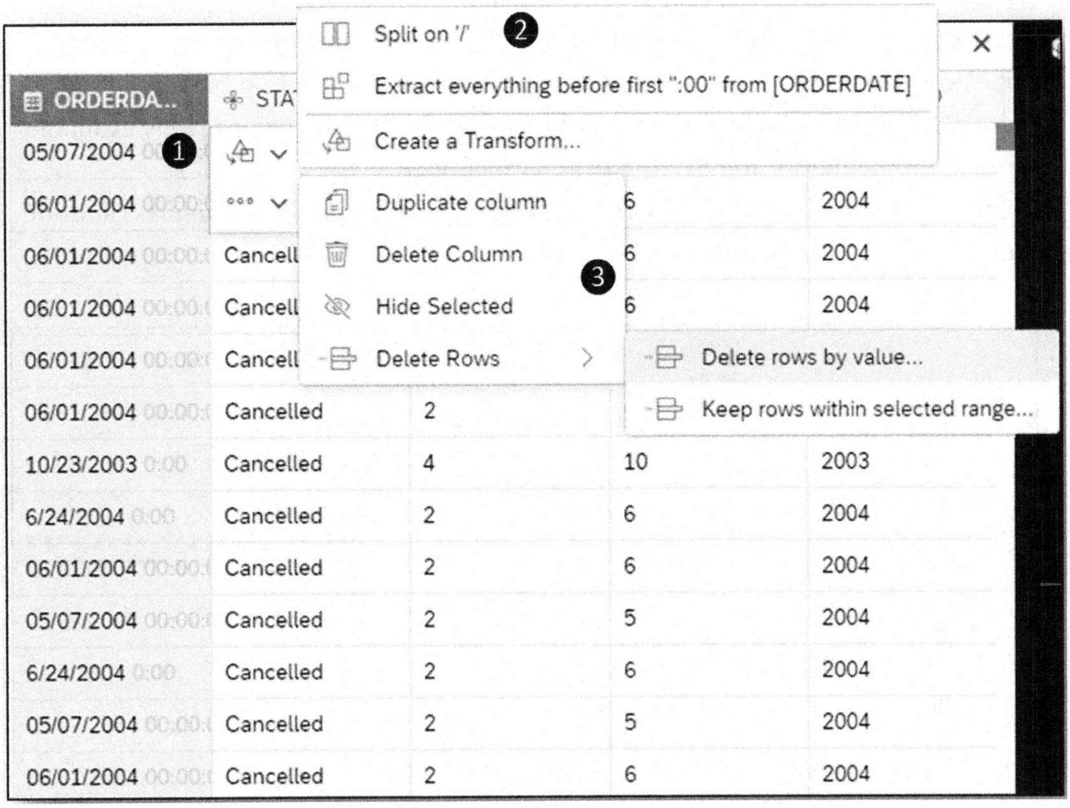

Figure 3-23. *Create Transformations in a Model*

Step 1: Click on the column where the transformation has to be done as shown in Figure 3-23.

Step 2: Click on the ⭤ icon as shown to create a transform. The **"Split"** transform can be used to spit the data in a column. The **"Duplicate"** transform can be used to quickly copy the contents of a column into a new one.

Step 3: Likewise, the **"Delete Column"** can be used to delete a particular column. Hide Selected can be used to hide columns and finally Delete Rows can be used to delete a particular set of rows from a column.

7. **Geo Enrich:**

Geo enrichment is the process of contextualizing data with geo dimensions. Once dimensions are geo enriched, they can be placed on a Geo Map for data exploration within a map. Geo Maps are visually more appealing and easier to decipher than raw tabular data. Dimensions can be geo enriched by locations data that is readily available within SAC or manually through latitude and longitude.

Figure 3-24 explains the process of enriching dimensions with coordinates (latitude and longitude).

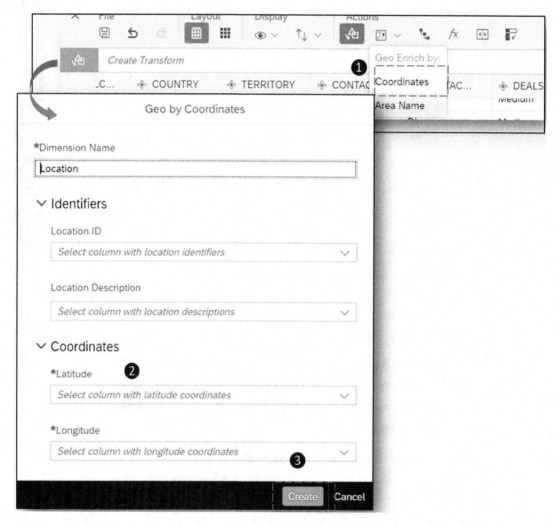

Figure 3-24. *Geo Enrich Dimensions by Coordinates*

Step 1: Select the "**Geo Enrich**" option and select the dimension
on which to Geo Enrich.

Step 2: Select the column that holds the latitude and longitude
information. Let us select the **"Country"** column.

Step 3: Click "**Create**" to create the Geo Enriched dimension.

Likewise, Dimensions can be geo enriched with Area Name already provided within
SAC, as shown in Figure 3-25. However, this information is limited to the United States
and does not allow drill down past city level. In order to make the most of geo maps, it is
advisable to have coordinates data available.

Step 1: Select the Dimension name, which in this case is **"Area."**
This is shown in Figure 3-25.

Step 2: Select the column on which to build the Geo Enrichment,
which in this case is **"Country."**

Step 3: Create the Geo Enrichment.

Figure 3-25. *Geo Enrich Dimensions by area*

8. **Level-Based Hierarchy:**

This option allows users to create a hierarchy in the model. We discussed hierarchies in detail in the section **"Models"** and subsection **"Hierarchies"**. Figure 3-26 shows the option to create a level-based hierarchy.

Figure 3-26. *Create level-based hierarchy*

Step 1: Click on the + symbol to add a level to the hierarchy. Name the hierarchy as "**LocationHierarchy**" as shown in Figure 3-26.

Step 2: Add columns to the hierarchy and click "**OK.**" We have added **Country, City,** and **State.**

Step 3: A pop up will inform how the hierarchy will be structured.

The hierarchy is visible in the model properties.

9. **Calculated columns:**

A calculated column is a new column that can be created using the existing columns. Source data might not always be aligned to the requirements of the model. Also, models enable data enrichment through transformations and cleansing capabilities. However, if the requirement is to have data that cannot be achieved through the available transformations, calculated columns have to be created in the model, which enable advanced data enrichment through formulas. The calculated column screen consists of three components as shown in Figure 3-27, which are the following:

- Edit Formula for creating and editing formulas as shown in ❶

- Formula Functions for the list of available formulas as shown in ❷

- Preview for previewing the results of the calculation as shown in ❸

Let us now try to learn about calculated columns with a scenario. Consider the source data has two columns for address line, **ADDRESSLINE1** and **ADDRESSLINE2**. The requirement is to have a single column for address line, **New_Address**. Let us learn the process of creating a calculated column for this scenario with a step-by-step process.

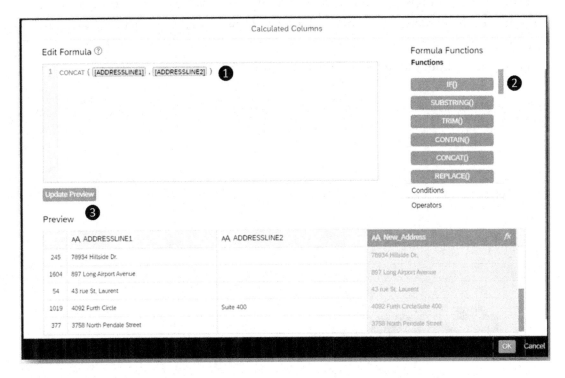

Figure 3-27. *Calculated Columns*

Step 1: The Formula window is used to build the formula for the calculated column. Build the formula using the functions available. For merging the two address lines, we use the **"CONCAT"** function for concatenation. This is shown in Figure 3-27.

Step 2: The Formula Functions bring up the functions available for creating the calculated column. We have selected the **"CONCAT"** function.

Step 3: On building the formula, check the validity in the preview window. Then **New_Address** column for concatenated address.

Step 4: Click "**OK**" to add the calculated column to the model.

10. **Combine Data:**

This is the most important aspect of the requirement of Single Version of Truth. Data from multiple data sources can be combined into a single model to provide end users with a single unified data source for analysis. Let us learn about how data can be combined into a single model.

Data can be combined across diverse data sources as well as files. In this particular example, we will learn how to combine data from another file into the model we have created for sales.

Let us learn about the steps as shown in Figure 3-28.

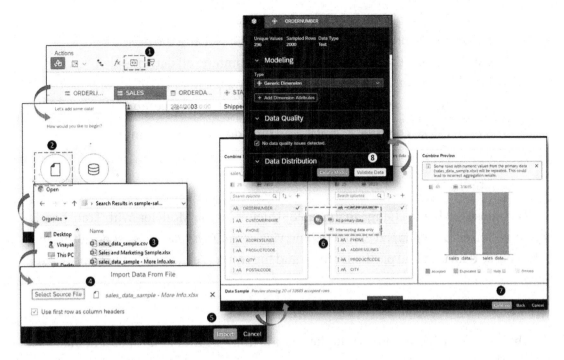

Figure 3-28. *Combine multiple datasets*

Step 1: Select the option to **"Combine Data"** as shown in Figure 3-28.

Step 2: SAC allows data from a file or an existing data source to be combined into a single model. In our example, let us combine data from a new file **"sales_data_sample – More Info.xlsx".**

Step 3: The **File Explorer** opens up. Select the file for merging with the data already existing in the model, which is **"Sales_data_ sample".**

Step 4: SAC shows a confirmation of the file selected. We have in our case provided additional information to the already existing data in the model.

Step 5: On clicking "**Import,**" data is imported into the SAC platform as shown in Figure 3-28.

Step 6: The "**Combine Data**" Window comes up. This window allows combining data from the existing dataset and the new dataset. Select the option on whether to combine all data or only intersecting data. The adjacent graph shows the data quality of the data combined and how accurately the data has been combined.

Step 7: Click "**Combine**" to finish and include the new data into the data model.

11. **Transpose:**

Transpose enables a change to the table structure in the Modeler. With Transpose, columns can be interchanged with rows and vice versa. This feature is especially useful if the source data is in one particular form and the result is expected in the transposed form.

12. **Remove Duplicate Rows:**

This option enables removing duplicate rows from the available dataset in the Model. This is a crucial feature in data cleansing, which enables automated cleansing of duplicate data saving considerable time for end users.

This covers our learning of the Top Menu of the Modeler. Now let us learn about the Side Menu of the Modeler.

Side Menu Components

We will now discuss the Side Menu for the Modeler. While the Top menu is especially useful for creating the model, the side menu bar is essential for refining the model parameters and finalizing a "Single Version of Truth" for end users to explore. The Side Menu is shown in Figure 3-29. Let us learn each of the components of the Side Menu and how they contribute to Model creation and design.

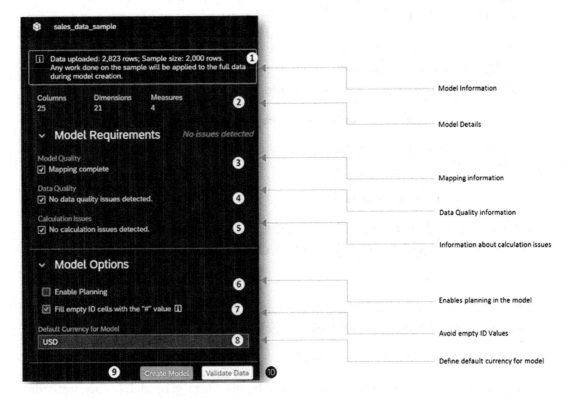

Figure 3-29. *Model Details Menu*

1. **Model Information:**

This section provides a summary of the entire model, which proves to be a great reference for developers. This section also shows how many rows of data were uploaded and the size of the sample dataset on which is considered during development of the model.

2. **Model Details:**

Further details on the number of dimensions and measures are displayed in this section. This section also shows how many columns have been uploaded with the data, and how many of these have been classified as dimensions and measures from the dataset.

3. **Mapping Information:**

This section provides information on the mapping of the columns. If there are issues found with the mapping, this section will bring out these errors.

4. **Data Quality Information:**

The next section brings out data quality errors. In the current example, there are no data quality issues. However, in case there are any, this section will highlight them in red. These issues need to be fixed in order for the model to be created.

5. **Information about calculation issues:**

Likewise, the issues in any calculations created in the model are highlighted in this section.

6. **Enable Planning in Model:**

If a planning model is being built, ensure that this option is clicked. Planning models are different from BI models and will be explored further in subsequent sections.

7. **Avoid empty ID values:**

ID values act as keys in ensuring that the data model is consistent with the data values. If the values are allowed to be empty, click on this option.

8. **Define Default currency in Model:**

The final option is for the default currency to be used for the model. In this particular case we have used USD.

9. **Create Model:**

Once all the data loading, blending, and transformations have been completed, this option is used to finally Create the Model.

10. **Validate Data:**

Before creating the model, data needs to be validated by clicking on the **"Validate Data"** option. This option enables end-to-end data validation post the data transformations as well as fixing data quality issues. The SAC internal algorithm scans each value in the complete dataset and points out any outliers that do not fit in.

The above view for the Side Menu as well as the Top Menu is for the entire model. However, details of a particular column can also be viewed in a very similar screen, which is shown in Figure 3-30. This view has similar options as that of the model but is restricted to only one column and has details specific to that column only.

Figure 3-30. Column details menu

Thus, a model can be used to combine data from multiple sources through connections and present a "Single Version of Truth" to end users.

Model Preferences

Model Preferences allow end users to fine-tune the created Model with additional parameters. The parameters enable users with design-time parameters and are some of the features that can be changed even after the Model is created. Let us learn about the settings available in Model Preferences and how they aid in fine-tuning the created model. The Model Preferences are as shown in Figure 3-31.

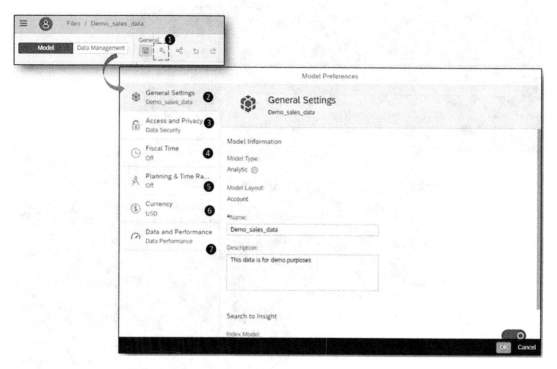

Figure 3-31. *Model Preferences*

    1. **Access:**

Model Preferences can be accessed by clicking on the ✎ Icon from the Top Menu. The Model Preferences window comes up as shown in Figure 3-31.

    2. **General Settings:**

The General Settings include settings for the type of the model (analytics or planning), name, and the option to enable indexing for Search to Insight. We will learn about Search to Insight in further detail in Chapter 5, when we learn about Augmented Analytics in SAC.

3. **Access and Privacy:**

The Access and Privacy settings enable defining user security and auditing for the Model. This settings option enables and also displays the privileges available to the user with respect to the model.

4. **Fiscal Time:**

Fiscal Time denotes the financial year that is followed for the organization or the region. The Fiscal Time option enables defining the fiscal year with respect to the start and end month as well as the unit for the time, either as Period or Month. This option is especially useful for planning models where planning is done with the fiscal year as the base for calculations.

5. **Planning and Time Range:**

The Planning and Time Range option is enabled if the model is defined as a Planning Model. This option also shows if there are any private versions associated with the particular planning model. A version is a copy of the planning data, which can be private when stored for an end user and public if stored for all users.

6. **Currency:**

The Currency option allows setting up the default currency for the Model.

7. **Data and Performance:**

Enabling this option enables better performance in the model by optimizing updates when changes are made to the chart or table.

Summary

In this chapter, we have learned in depth the step-by-step process to create connections and models. We have further learned about the Modeler in SAC, which is used to create models and all the options that it has to offer. We have combined the inputs from multiple data sources into connections and combined them into models to deliver "Single Version of Truth" to enable uniform data and better decision-making.

This brings us to the conclusion of the chapter on how SAC enables a "Single version of Truth" across the enterprise. We learned about the following:

1. The importance of having a reliable single version of true data for reporting and analysis.

2. Step-by-step process to create connections and models.

3. SAC data modeler tool.

In the next chapter, we will learn how SAC will be able to meet another significant requirement of ABC Inc., which is "the All-in-One" Analytics Platform.

CHAPTER 4

Leverage SAC to Create "All-in-One" Analytics Platform"

Being a large global enterprise, ABC Inc. has multiple business functions cutting across Lines of Business. Each of these business functions works independently with its own set of tools. This has resulted in multiple tools being used across locations and business functions for the same or similar decision-making processes. Multiplicity of Systems compound the problem of data Redundancy – scattered across silos and systems within the enterprise IT landscape. The challenge becomes graver due to the requirement of disparate data sources used by support business functions like strategy and business planning. For example, planners within the organization would be using a specific tool for their day-to-day activities. Their reference or base dataset would be curated solely for their purpose. There would be data scientists who use external data sources. A single set of curated data is specific to a purpose and seldom gets reused. The Enterprise Datawarehouse has over the years grown exponentially on which smaller data marts have been built, mostly with a lot of data redundancy. This data has come at the cost of increased storage and processing capability. ABC Inc. currently has an Enterprise Data Warehouse; however, business users tend to build their own sets of data locally or across shared drives. Not only does this lead to data discrepancies, but also divergencies in decisions across business functions. For instance, when top leadership wants to review actual vs. planned, there are regular instances of discrepancies due to diverse datasets. Different solutions and technologies for decision support also lack standardization in terms of references; for instance, there are differences in calculation of KPI or their definitions across units within the same line of business.

© Vinayak Gole, Shreekant Shiralkar 2020
V. Gole and S. Shiralkar, *Empower Decision Makers with SAP Analytics Cloud*,
https://doi.org/10.1007/978-1-4842-6097-5_4

Lack of Collaboration

Multiplicity of Business Planning and Decision Support systems hinder critical work aspects like collaboration and remain the main causes for delays and high inefficiencies at ABC Inc. Sales staff at ABC Inc. spend a lot of time in the office analyzing data instead of focusing on the market; they have to then collect and collate data to prepare and validate the data over email with their sales managers, who in turn do so with their national managers to get a quick summary dashboard for global management to get a sense of business transacted on the previous day. Similarly, the finance analysts and other analysts remain engaged in the nitty-gritty of the data by looking at the previous day's sales, what products are being sold, and where they're being bought, instead of producing insights that enable decisions for higher impact to the business being transacted or shape the business to be transacted.

TCO is high due to management of multiple applications apart from a host of different vendors and suppliers, partners, and also a labyrinth of Hardware and Software Solutions. Moreover, tools have specific hardware requirements and ABC Inc. has to invest and engage multiple resources to maintain these servers. These tools have to be upgraded from time to time, which in itself has been a time-consuming and complex procedure.

In the previous chapter, we analyzed the requirement of having a single version of truth from data across data sources. Additionally, multiple business functions have to collaborate regularly for optimal functioning. In the absence of a robust collaboration tool, data governance and security become issues. The collaboration across functions is currently through enterprise collaboration tools. This is cumbersome and difficult to track, resulting in miscommunication if not missed communication across the functions. It is well recognized that having an integrated platform for all types of analytics is equally essential for realizing a data-driven culture.

In view of the current situation, the requirement is to have a single platform wherein all analytics, both basic and auxiliary, can be made available to business users. ABC Inc. also wants to reduce operational costs by moving this platform to the cloud or opting for a SaaS software as a service-based product. This platform should enable primarily a 360-degree view of enterprise data, self-service for business users, and planning functions for enterprise planners. Custom application development and predictive analytics are desirable but not in the list of primary requirements.

ABC Inc. desires a solution that enables data-driven decision-making a reality, in a way that someone in Melbourne and somebody in Texas in a meeting refers to and uses, the same tool, looking at the same data in real time. Additionally, the platform should enable robust data governance and security to ensure the right data is available to the correct set of users. End users across locations and business functions should be able to collaborate across a single platform with history tracking and versioning to make it easier to make the changes. The platform should also encourage Agile BI and be compatible to continuous deployment in tune with current industry trends.

Alignment to Specific SAP Analytics Cloud Capability

With the expansion of the analytics portfolio and demand to consolidate multiple analytical systems into a single platform has been gaining steam. Analytics is no longer considered just a back-end system, but now forms the core pillar in SAP's Intelligent Enterprise framework. Holistically, Analytics forms the driving force behind decisions being driven by data rather than on hunches and experience. And if these analytic processes can be made available on a single platform, end users can reduce the effort of logging on to multiple systems and consolidating data. And one of the major advantages would be reduction of human errors in the reconciliation.

SAP's design view for SAC's architecture has been to promote a single platform based on HANA to encompass analytics, planning, and predictive analytics. SAC is also the focus for SAP innovation and sees regular updates in terms of new technologies being rapidly incorporated into the platform. SAC is also native to the cloud, built from scratch on the SAP cloud platform, which enables access to analytical functions from multiple devices. These features make SAC the tool of choice for bringing all analytics to a single platform.

Some of the benefits that SAC brings about for bringing all analytics to a single platform are the following:

1. **HANA as the foundational platform:**

   SAC is built on the HANA Cloud platform. Having a versatile platform on the cloud with a proven database technology enables SAC to be equally flexible in terms of approaching analytics holistically. The HANA platform brings to the table robust data processing skills and new age storage technologies. Building on this framework, SAC delivers a complete packaged set of analytics functions that can be based out of a single data processing layer.

End users can exploit the rapid data processing capabilities of HANA as well as other features of the platform for real-time analytics as well as faster insights to action.

2. **Faster insight to action:**

Having integrated disparate system functionalities, it becomes imperative to have good collaboration. SAC's robust collaboration tools between end users and connectivity between tools ensure less time is spent on discussions and analysis of results. Decisions can be arrived at quickly with traceability within the system itself. Time saved in terms of sifting through emails and setting up meetings can be reduced and constructively spent on activities adding value to the organization.

3. **Lower Cost of Ownership:**

As we have already discussed, one of the innate qualities of SAC is its capacity to integrate multiple functions related to analytics onto a single platform. If HANA provides the data processing capabilities, the application layer of SAC brings together disparate functions that would otherwise require multiple systems to process and use. For example, planning, predictive, and analytical datasets can be combined to bring out stories that reflect a 360-degree view of the data. The singular SAC platform caters not only to end users in day-to-day self-service data analysis but also the top-level executives to explore data over the Digital Boardroom in real time. Developers can build custom applications whereas planners can plan and share data collaboratively. And moreover, data scientists can bring in predictive functions to put out a road map for the next steps and the road ahead.

One of the primary benefits of a single platform is the total cost to the organization in terms of license and infrastructure. SAC's SaaS billing options ensure there is rapid scalability available as and when needed.

In the last chapter, we learned how models are the base for bringing data across sources and consolidating it into a single true version of data. In this chapter, we will learn the components that enable SAC to deliver multiple analytics products over a single platform. The components are the following:

- Stories

- Planning Capabilities

- Digital Boardroom

Next, let us learn about Stories in detail.

Stories

Reports represent data in the form of tables and graphs. Stories, however, bring in the element of storytelling to data. Stories that are presentation-style documents are more effective in bringing out contextual information of the data. Stories bring data to life, enabling end users to discover hidden meaning deep into the raw data.

Stories can be one of these types:

- Report

- Dashboard

Let us learn about each type of Story in detail.

Report

One of the most widely accepted formats of representing data is in the form of tables and graphs. Reports are widely static and do not allow for a lot of flexibility in terms of self-service and data analysis.

Dashboard

Dashboards offer a bird's-eye view of select KPIs across the organization. Dashboards typically help senior management get a consolidated view of the entire enterprise. Dashboards also enable business functions to have regular updates on how their lines of business are doing. Dashboards are always at an aggregated level whereas reports are at a detailed level.

Planning Capabilities

Planning is an essential business function that enables enterprises to define a path toward achieving predetermined goals. Planning is a specialized function and requires an equally unique set of tools to determine plans, budgets, and forecasts for enterprise functions.

Planning is an integrated component in SAC that integrates with other SAP planning tools and ERP systems to deliver a unified planning experience in the cloud. SAC planning tools are comprised primarily of planning models.

Planning Models

While analytic functions enable end users explore and report on data, planning models enable end users to develop budgets and forecasts. Planning functions always have a time dimension to enable planning features like spreading, distribution, and allocation.

Planning models have all the capabilities of analytical models but also enable planning functions as described earlier.

Value Driver Trees

In simple terms, a value diver tree is a what-if simulation that enables end users to simulate the outcome of their decisions. Value driver trees enable end users to try out combinations of simulations across interdependencies of measures.

For a value driver tree, the underlying model must be a planning model and have hierarchies defined, which would enable simulations across interdependencies.

Planning Functions

Allocating, spreading, and distribution are the primary planning functions that enable end users to build budgets and forecasts. SAC allows end users to create private as well as public versions that can be used across the enterprise. Workflows enable review and finalization of the final plan, which is then used as a baseline for plans throughout the year.

Digital Boardroom

Enabling the boardroom with real-time analytics, Digital Boardroom is one of the differentiators of SAP Analytics Cloud. Enabling executives to get a 360-degree overview of the organization, the Digital Boardroom experience is delivered over a single or multiple touch screen. The boardroom enables setting up of agendas and presentations and enables real-time drilling up, down, and across to drastically reduce the time taken to deliver information and answers to business leaders.

The Digital Boardroom experience works over and above the SAC platform and enables interlinking stories, planning, and predictive pieces to create a live presentation that enables real-time contextual information delivery. Businesses can arrive at critical decisions over a single meeting without waiting for data to be made available over multiple requests.

In the next section we will learn the step-by-step process how to build each of the above components and integrate them with stories that we have learned about previously.

Enabling an "All-in-One" Analytics Platform

We discussed how not just Analytics but also supporting business functions can be delivered over a single platform with SAC. In this section we will learn the step-by-step process to build the features we learned about in the previous section.

Note   Since SAC is a constantly evolving product with SAP introducing new features every quarter, some of the actual screens might be different from what are shown in this book.

Creating a Story

A story is where data is visualized and explored with a combination of visual elements including charts, graphs, and tables. A story can have a single or multiple pages that can be used to build stories on the underlying dataset. A story can be built using the Story Builder interface in SAC. Follow the steps below to start the story builder interface.

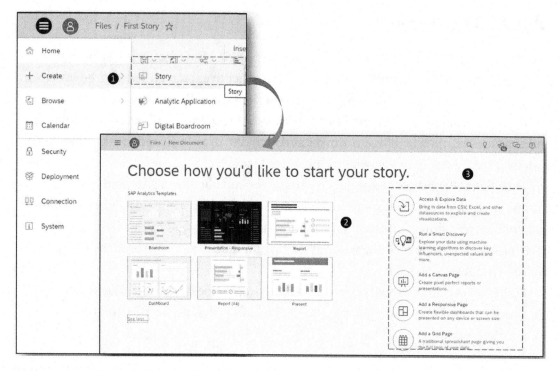

Figure 4-1. *Create Story Options*

Click on the main menu as shown in Figure 4-1 and click on **Create** option and then on **Story**. This brings up the Story Builder Wizard interface as shown in ❶. The Story Builder Wizard provides multiple options to create a story. The core components of the Wizard are discussed next.

Templates

This component provides ready-to-use templates for different types of stories. Ready-to-use templates are available for creating Dashboards, Presentations, and Reports. We have learned about dashboards, presentations, and reports in the previous section on "Alignment to Specific SAP Analytics Cloud Capability." Depending on the type of story being created, a template can be selected and used. This saves considerable time in formatting and building the presentation of the story. The available templates are shown in Figure 4-1 ❷.

Story Creation Options

The Wizard also provides multiple options to select how to create a story. The options available for creating a story are listed as below and are shown in Figure 4-1 under ❸. We will learn about each of these options in the section "Step-by-Step Process to Build Interactive Stories."

- Access and Explore Data

- Smart Discovery

- Add a Canvas Page

- Add a Responsive Page

- Add a Grid Page

Selecting any of the above options brings up the Story Builder Interface. In the next section, let us learn about the components of the Story Builder Interface.

Story Builder

The Story Builder enables building stories from underlying models or directly on datasets. Before we learn the step-by-step process to create a story, let us learn the components of the Story Builder Interface.

The components of the story builder interface are the following:

- Story tab

- Data Tab

Note   For details on the Story Builder, refer to Appendix A.

Canvas

The canvas is shown in Figure 4-2. It can be styled with blocks as shown. In the current example, a chart has been placed in a single block. The canvas allows the addition of tabs/pages as shown in ❶ and placing of an object as shown in ❷.

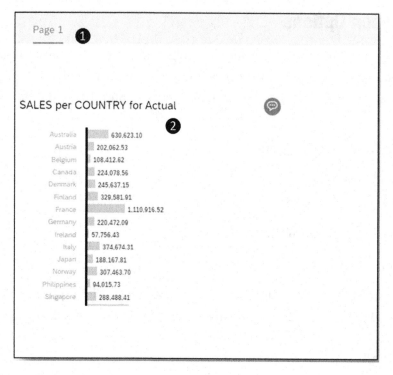

Figure 4-2. *Story Canvas*

The Story Builder Menu

The Story Builder Menu contains options and settings for the story as well as story components and widgets. These options are essential for building and then stylizing the story.

Note   For details on the Story Builder Menu, refer to Appendix A.

Building Interactive Stories

SAC provides multiple options to build stories either directly through a data source, file, or a data model, and we shall learn each of these options in detail in the coming section.

Stories can be built as shown in Figure 4-1 in options 1 to 5 below. Click on the **create** option and select **Story**. This will bring up the **Story Creation Wizard,** which offers multiple options to create a story.

This section provides Five Options for creating stories. We shall learn each of these options in detail.

Option 1: Access and Explore Data

In this option, SAC allows us to create a story by exploring a dataset. The Access and Explore option enables creating a Story from the following:

- Data uploaded from a file

- Data from a data source

- Data acquired from an existing dataset or a model

Create Story from Data Uploaded from a File

SAC allows stories to be created by simply uploading files from the local machine or a shared drive into the cloud platform. In this particular example, we shall learn how stories can be created by exporting a csv file from the local file to the SAC platform.

Figure 4-3. *Create Story from data uploaded from file (Steps 1-6)*

Step 1: Click on the first option, which is to **"Access and Explore data."** This is shown in Figure 4-3.

Step 2: The **"select data"** window comes up. Select the first option, which is to select a **"data uploaded from a file"** as a data source.

Step 3: The select file option comes up. Select the source file by clicking on the button as shown in Figure 4-3.

Step 4: Since we are uploading a local file, the file explorer comes up. Select the file that you need to be explored. In this particular case, it is the sales_data_sample.csv.

Step 5: Since we have selected a csv file, the SAC needs to understand the delimiter, which needs to be specified. Select the delimiter as shown. Also mark the first row as column header if the first row in the source file is a header. Click "**Import**" as shown in Figure 4-3.

Step 6: SAC displays a message on information about the dataset, including the sample size, which in this case is 2000.

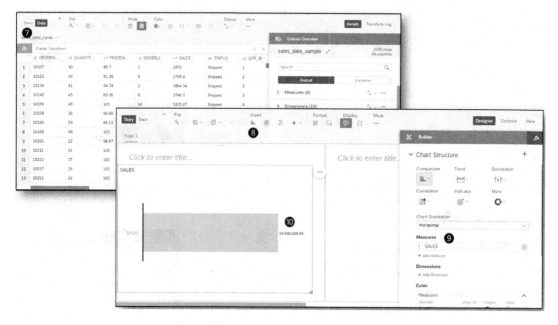

Figure 4-4. *Create Story from data uploaded from file (Steps 7-10)*

Step 7: The **Data Tab** also enables making changes to the dataset. Once the changes are made, click on the Story Tab. This is shown in Figure 4-4.

This view is similar to the modeler we learned about in the previous chapter. However, a separate model is not created, and the data imported is only for this particular story.

Step 8: On the **Story Tab**, click on **"Insert"** as shown in Figure 4-4. A chart is inserted into the canvas.

Step 9: Select the Dimensions and Measures from the Story Builder Menu. We have selected the measure **"Sales."**

Step 10: The chart is created as shown in Figure 4-4. Further components can be added as above to create the complete story. Click on **"Save"** to save the story. We will save the story as **"Sales_demo".**

Create Story from Data from Data Source

SAC also allows for data to be created directly from a data source. Using a connection and sources exposed from external systems, SAC allows stories to be built on the sources.

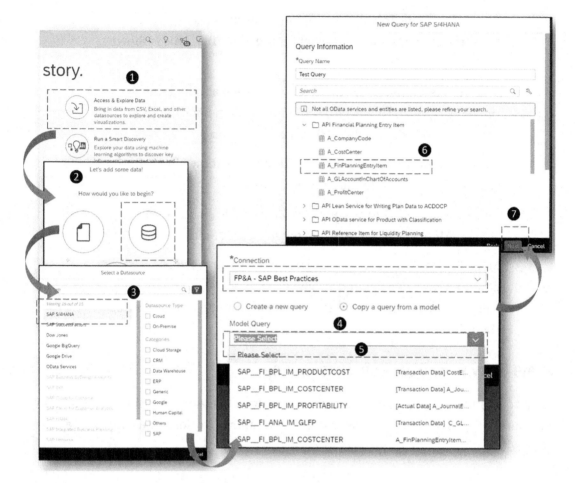

Figure 4-5. *Create Story from Data from Data Source (Steps 1–7)*

Step 1: Click on "**Access and Explore Data**" as shown in
Figure 4-5.

Step 2: Select the **"Data from a datasource"** option from the
"Data selection" window.

Step 3: The list of all data sources available within the SAC
landscape come up. This is the same option that comes up while
creating a connection. Select the type of data to connect to.

Step 4: We would be connecting to SAP S/4HANA in this particular
example. Select S/4HANA as the data source to connect to.

Step 5: The next window brings up the connection information. All the connections that have been created with S/4HANA as source are shown. Select the appropriate connection as shown in Figure 4-5. Once the connection is selected, the next option is to select the query that would act as the data source for the story.

Select from the option to either **"create a new query"** or to **"select from an existing model."**

Note   This is one of the advantages of having a model since it can be reused and acts as a semantic layer.

In this particular example, let us **"select from an existing model"** option.

Step 6: In the next window, name the query and select the CDS view that has been exposed as OData. CDS views or Core Data Services form the base of data management in S/4HANA. For consumption in SAC, the CDS Views must be exposed as OData (@OData.publish:true).

Step 7: Click "**Next**" as shown in Figure 4-5.

Figure 4-6. *Create Story from Data from Data Source (Steps 8–11)*

Step 8: The components of the CDS views are displayed. The already existing query is displayed as shown in Figure 4-6. The query can either be edited or used as is.

Step 9: Click "**Create.**"

Step 10: The modeling interface comes up. Transform the data as and if needed.

Step 11: Click on the story tab, which brings up the designer for creating the story. Select the measures and dimensions from the Builder.

Step 12: Create components of the story as shown in Figure 4-6. Click on **"Save"** to save the story. We will save the story as **"Sales_ demo".**

Create Story from Data Acquired from a Dataset or Model

Models enable the concept of semantic isolation as well as building a layer for data management in SAC. We have already learned how models can be built and data transformed in Chapter 3.

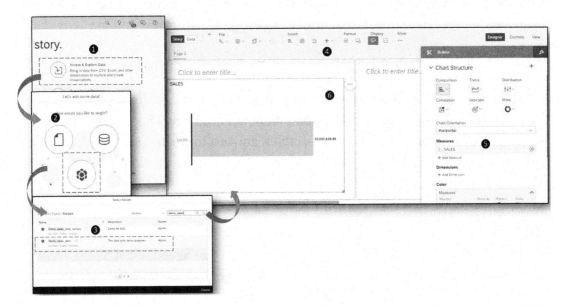

Figure 4-7. *Create Story from Data acquired from an existing dataset or model*

Step 1: Click on "**Access & Explore Data**" in Figure 4-7.

Step 2: The Data selection window comes up. Click the option for selecting "**data acquired from an existing dataset or model.**"

Step 3: The "**Select Model**" window comes up. In this window, select the model on which the story needs to be built.

Step 4: Click on the "**Insert**" button and select the chart to be placed within the story.

Step 5: In the **Story Builder**, select the dimension and the measure to include in the story as shown in Figure 4-7. We have selected "**Sales**" for creating the chart.

Step 6: Build the story with charts, tables, and other objects. This is as shown in Figure 4-7. Click on "**Save**" to save the story. We will save the story as "**Sales_demo**".

Option 2: Smart Discovery

SAC provides multiple augmented analytics features, the prominent one being Smart Discovery. Smart Discovery allows stories to be created, aided by underlying Machine Learning features of the SAC. Smart Discovery identifies the key influencers and relationships of data. We will cover this part in more detail in Chapter 5 when we discuss augmented analytics within SAC. For now, let us move on to the next option.

Option 3: Add Canvas Page

Traditionally a canvas is where a painting comes to life. In SAC, a canvas is referred to the blank page that acts as a starting point for building a story. This page can be molded to include images, charts, graphs, tables, and text to create a story exploring data contextually. See Figure 4-8.

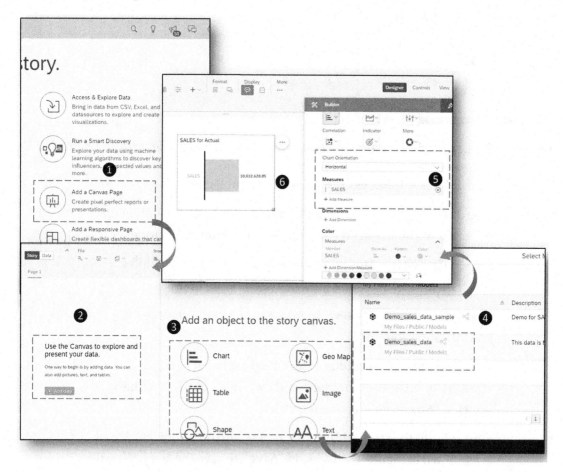

Figure 4-8. *Add canvas page*

A canvas page is ideal for pixel-perfect standard dashboards with defined dimensions. If viewed in lower resolutions, scroll bars appear but the layout does not change. A canvas page allows more flexibility in terms of overlaying objects, grouping objects, or moving objects to the front or back. However, a Canvas page needs to be redesigned to suit different devices like laptop/desktop, iPhone, or iPad.

Here is the step-by-step process to create a story based on a canvas page:

Step 1: Select the option to "**Add a canvas page**" as shown in Figure 4-8.

Step 2: The next window offers options for data to be included into the story as shown under ❶.

Step 3: The other option is to directly include a story object as shown under ❷. Select the appropriate option to include in the story.

Step 4: On selecting either of the options, the model selection window comes up.

Select the model as shown in Figure 4-8. We have selected the model we have already created in Chapter 3, which is **"Demo_sales_data"**.

Step 5: Select the dimensions and the measures from the story builder. We have selected the **"Sales"** measure.

Step 6: The chart is displayed as shown in Figure 4-8. Additional objects can be added to the page to create a pixel-perfect dashboard. Click on **"Save"** to save the story. We will save the story as **"Sales_demo".**

Option 4: Add a Responsive Page

A responsive page is capable of adjusting the story as per the dimensions of the device on which a story is rendered. A responsive page is especially useful if the story is being used across multiple devices like computer screens, mobiles, tablets, etc. With mobility gaining a precedence in recent times with the feature to access data anytime anywhere, the responsive page proves to be a very useful feature. While viewing on smaller devices, data labels and legends are automatically hidden to increase the display area. The same story while being viewed on a larger screen is capable of spreading the information as per the screen real estate available. Responsiveness of the screen is as shown in Figure 4-9. As the story changes screens, it is adjusted per the available screen size.

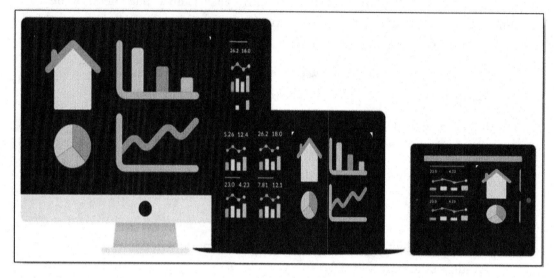

Figure 4-9. *Responsive Screens*

Lanes

Lanes provide the architecture for responsive pages to resize across screens of different sizes without any manual adjustment. The responsive canvas in SAC can be divided in up to six lanes to enable tiles placed within them to adjust accordingly. Each screen is categorized as Large, Medium, or Small. Depending on the screen size of the device on which the content is being viewed, lanes reflow content to fit the screen size. Widgets can be placed as tiles in each of the lanes and when the screen size changes, the lanes adjust to fit into the available screen size. The other lanes can be accessed by either scrolling down or scrolling across depending on the type of screen. Though lanes provide flexibility in terms of responsiveness, they come with certain drawbacks. Overlaying of widgets in tiles is one such drawback. Each widget has to be placed as a distinct tile and cannot by overlaid. Lanes thus provide the much-needed flexibility to the story.

Let us now learn the step-by-step process to build a story with a responsive screen layout. The entire process is shown in Figure 4-10.

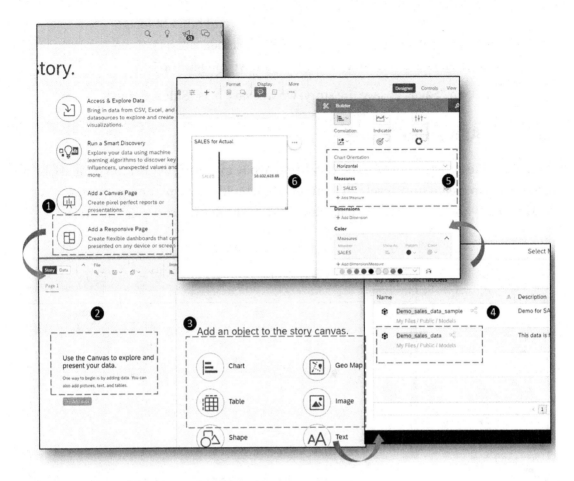

Figure 4-10. *Add Responsive Page*

Step 1: From the Story Builder Interface shown in Figure 4-10, click on the option "**Add a responsive page.**"

Step 2: Select the data to be uploaded as shown under ❷.

Step 3: Or select the object to be placed on the canvas under ❸.

Step 4: Select the model as shown in Figure 4-10. We have selected the model we already created in Chapter 3, which is "**Demo_sales_data**".

Step 5: Select the dimensions and the measures to be used in the story. We have selected only the **"Sales"** measure.

Step 6: Create the story using graphs, charts, and other objects. This story will be able to adapt itself to the screen size as per the device on which it is viewed, which is shown in Figure 4-10.

Click on **"Save"** to save the story; we will save it as **"Sales_demo".**

Option 5: Add a Grid Page

A grid is a table structure and can be very useful while exploring detail level data. It can also be used to create story level calculations while building the planning strategy.

Let's learn how to build a grid page into a story. The entire process is shown in Figure 4-11.

Figure 4-11. *Add Grid Page*

Step 1: Click on the option to **"Add a Grid Page"** as shown in Figure 4-11.

Step 2: An empty grid comes up. This is the grid page.

Step 3: Click on **"Add"** under **"Insert."** This brings up the Model Selection window.

Step 4: Select the appropriate model to build the Grid Page upon. We have selected the model we have already created in Chapter 3, which is **"Demo_sales_data"**.

Step 5: Select the measures and dimensions from the story builder to build into the grid page. We have selected **"Country"** and **"Amount"** as shown in Figure 4-11.

Step 6: The Grid page is created as shown. It can be augmented further with calculations and data transformations. With all the qualities of data manipulation of a spreadsheet, the grid page can be a formidable tool for the data analyst.

We have learned how to create a story from multiple types of sources and display on multiple pages. We will now learn about some of the unique features of a Story.

Story Features

A SAC story not only consists of components that can be added to the canvas but also features some very unique features that further enhance the capabilities of the story. These features enable data analysts to delve deeper into data, designers to build better dashboards, and end users to better arrive at decisions. We will now learn about some of these features.

Link Models

In Chapter 3, we learned how data from multiple sources can be blended into a single model to create a consolidated dataset for further exploration. Data thus collected into a single model enables achieving "Single Value of Truth" across the enterprise.

Architecturally sound design dictates using models based on lines of business and use cases. SAC also enables bringing data across multiple models into a single story. This functionality enables a further level of aggregation and blending of data stored in models into the story. Organizations typically have to blend data across different lines of business to get a 360-degree view of data. For example, in ABC Inc., models can be built across Sales and Finance. For gauging the impact of drop-in sales on the financials, the Sales model and the Finance model would need to be linked and a story built on the linked model. Linked models are especially useful for business analysts and organizational leaders to access the change in a parameter and its impact on the entire organization.

Models can be linked by linking common dimensions. A common dimension is essential to building a linked model to ensure data integrity across the models being linked. If there is no common dimension and the data is incorrectly linked, then the linked model will put forth incorrect values in the subsequent stories. Continuing on our previous example at ABC Inc., for linking a Sales model with a Finance model, a possible common dimension could be the Sales Organization. A common link of the Sales Organization can enable blending data across both models of Sales and Finance, enabling creation of a linked model and subsequent story.

Let us now learn the step-by-step process to link two models. The process is shown in Figure 4-12.

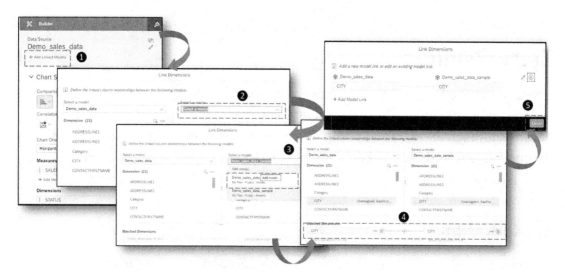

Figure 4-12. *Link Models*

> Step 1: Click on the option to "**Add Linked Models**" as shown in Figure 4-12.

> Step 2: The **Link Models** window comes up. This window shows the current model and the option to add another model.

Step 3: Select the model as shown in Figure 4-12. In this particular case, we have considered two models: "**Demo_sales_data**" and "**Demo_Sales_data_sample**".

Step 4: Models should be linked based on common dimensions. In this particular case, we have selected **"City"** as the common dimension. SAC shows if the linking is correct.

Step 5: Click **"Done"** as shown in the Figure 4-12. Once the dimensions are linked, data from both the models appear as a single data source on which the story can be built.

Chart Properties

We have seen how charts can be added to a story to create stories that are more informative and interactive. Chart properties enable additional features to be enabled for charts and make them more attractive.

Note   For Details on the Chart Properties, refer to Appendix A.

Conditional Formatting

Conditional formatting is the option to change the format of cells or charts based on certain predefined conditions. Conditional formatting enables us to bring out certain data conditions, visually enabling end users to spot trends easily and bring out anomalies without spending time on going through multiple rows of data.

Thresholds define levels for acceptable values. If a certain value falls below or above the threshold value, it can be considered critical and deserving of further analysis and action. Let us now see how conditional formatting can be enabled to bring out the data values that fit in to predefined conditions. In this particular example, we will work with the measure sales. If sales cross over 200,000, it would require further analysis and hence has been displayed as Red. Similarly, sales between 100,000 and 200,000 also suggest minor scrutiny and hence have been brought out as Orange. All the other values are as expected and have been colored as Green. The entire process of setting up conditional formatting and thresholds is shown in Figure 4-13.

Figure 4-13. *Conditional Formatting*

Step 1: For enabling conditional formatting, click on "**Conditional Formatting**" from the main menu as shown in Figure 4-13.

Step 2: The **Conditional Formatting** window comes up.

Select the option to either add thresholds to the model or the story. In the story, thresholds can be added or colors can be assigned. In this example, we will learn how thresholds can be added to the story.

Step 3: The **Thresholds** window comes up as shown in Figure 4-13. Select the model and the measure.

Step 4: Define number ranges and assign the range to a particular color set. In this example, we have divided the available data into three sets:

- Sales between 0 to 100,000 as Green;

- Sales between 100,000 to 200,000 as Orange;

- Sales greater than 200,000 as Red.

Step 5: Click **Apply** as shown in Figure 4-13.

Step 6: From the story builder window, add the measure on which the threshold has been added. Select **"Sales"** for applying the threshold.

Step 7: Add the threshold that has been created in the previous step 4 to the measure selected.

Notice that the chart values change the colors accordingly. This is shown in Figure 4-13.

Custom Calculations

In certain cases, having global measures built in the model are not sufficient. Certain story-level requirements need calculations to be created in the story for consumption. Custom calculations are formulas that can be custom built within the story using functions that have been provided by the SAC platform. In the model we created in Chapter 3, we have a measure for Sales. The Sales measure can be used with other dimensions to create charts for trends or be displayed as a numeric point. However, to determine the number of products sold, the measure **"Quantity Sold"** would need to be created at the story level using the custom calculations features.

In this section, we will see the step-by-step process to build custom-calculated measures, which are a type of custom calculation. The entire process for building a custom measure is shown in Figure 4-14.

Figure 4-14. *Custom Calculations*

Step 1: From the Story Builder, click on the "**Measures**" option as shown in Figure 4-14.

Click on Calculations from the drop-down and add **"Create Calculation."**

Step 2: This brings up the Calculated Measures window, which has multiple options for creating multiple type of measures including calculated measure; restricted measure, which is a filter; and aggregation, which is used for summation and converting a dimension to a measure. Select the option for "**Calculated Measure.**"

The panel also shows the list of functions available for building the formula.

Step 3: Select the appropriate formula (Sales/Price of Each) for creating the measure " **Quantity Sold**." In this particular example, we have calculated the quantity sold by dividing the sales amount by the price of each quantity as shown in Figure 4-14.

This measure called **"Quantity Sold"** can now be used in the story as a regular measure; however, it would be available for consumption only within the limits of the story.

This completes our learning of creating Stories and how they enable one capability of the "All-in-One" Analytics platform by enabling Business Intelligence on the cloud. Let us now learn how to browse created Stories. Stories can be browsed from the Main Menu as shown in the Figure 4-15.

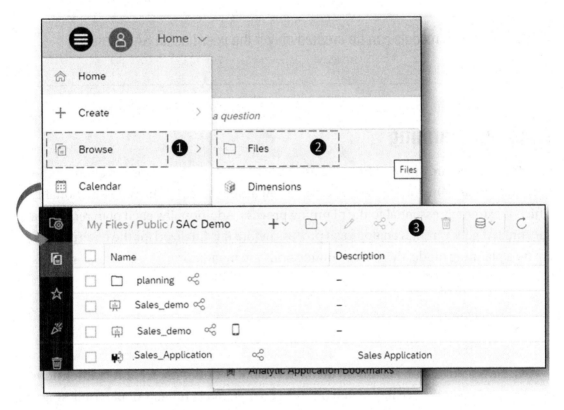

Figure 4-15. *Browse objects in SAC*

From the Main Menu, Click on "**Browse**" as shown in Figure 4-15 under ❶. The File Explorer comes up. Click on Folders to browse to files within the SAC Landscape as shown under ❷. The File Explorer can be used to browse for any object within the SAC including Models, Stories, and Applications as shown under ❸. We will learn more about folder structure and the File Explorer in Chapter 9.

We have learned how to create stories and browse them within the SAC landscape. Now we will learn about the next component, which is planning.

SAC for Planning

Business planning is necessitated by the need to set up budgets and plans and compare where the expenses stand at the current moment. Plans can be altered and allocations changed by using corporate planning functions.

SAC through its planning functionality enables the concept of all analytics in one platform by allowing analytics to work in tandem with planning functions.

In the next section, we will learn how to create a planning model step by step.

Note   Planning models can be created only if the user has an SAC Planning license.

Enabling Planning

Planning models are different from the analytical models we have learned about in the previous chapter. One of the cornerstones of a planning model is the time functionality, which is extremely essential to the planning process. Additionally, each plan requires a version that goes through an approval process before it is finalized for the enterprise. To create a planning mode, these three dimensions are mandatory:

- Category

- Account

- Time

We have seen the dimensions in detail in Chapter 3 under the section "Dimensions."

A planning model can be created on a live connection to other SAP planning systems like BPC. With a live connection to BPC, data can be fetched, modified, and written back to the source. However, SAC also enables planning with other sources like S/4HANA with an import connection wherein data can be sourced into SAC, planning processes implemented, and then written back to the source through jobs. We learned about Connections and Jobs in Chapter 3 in detail.

Create Planning Model

Like all other business processes, planning processes need to be collaborative, integrated, and automated. SAC provides both live connectivity to back-end planning systems like SAP Business Planning and Consolidation (BPC) as well as embedded planning in S4/HANA. SAC also provides the facility to upload data as files for carrying out the planning process. In the forthcoming sections, we will learn the step-by-step process of creating a planning model based on a file input for Actuals and Forecast Data. We will then build a story on the planning model created and compare the Actuals with the Forecast. Finally, we will learn about the Value Drive Tree and see how simulations can be brought in to the planning process.

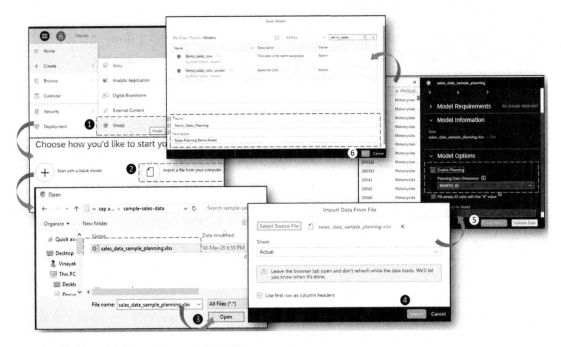

Figure 4-16. *Create Planning Model*

The first step for the planning model is to upload the Actuals data into the model. The entire process is shown in Figure 4-16

Step 1: Click on **CREATE** and then select **MODEL** for creating a model.

Step 2: Select the option to **"import a file from the computer"** as shown in Figure 4-16.

Step 3: Select the file from the local computer. Then click "**Open.**"

Step 4: Since in this particular example we will be loading the Actuals data first, select the Actual tab from the spreadsheet "**sales_data_sample_planning**". Click "**Import.**"

Step 5: The data is loaded into the model. Ensure that the "**Enable Planning**" option is selected. This enables the planning features of the model. Select the date dimension to baseline the time of the planning. In this example, we have selected the Month_ID. Complete the data transformation as we have seen in the analytical model. This is shown in Figure 4-16.

Click on "**Create Model.**"

Step 6: Give an appropriate name to the model and a description. We have named the model "**Demo_Sales_Planning**". The planning model is created. This model now consists only of the Actuals data.

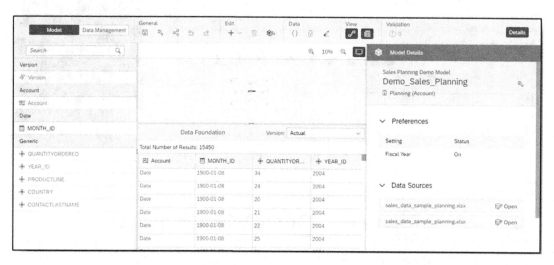

Figure 4-17. *Planning Model Structure*

Post creation of the model, the model structure will be as shown in Figure 4-17. Specifically, for planning models, there would be a version object that would hold the version of the changes being made. Currently, since only Actuals data is loaded, the version will show only one member. In the next section, we will add the Forecast data. Other members that could be added are Budget and Plan.

Figure 4-18. *Planning Model Load Data Jobs*

In the Data Management tab, it can be seen that an import job has been created. This is shown in Figure 4-18. This job will load the Actuals data from the Excel spreadsheet **"sales_data_sample_planning.xlsx"** from the Actual tab. This job can be scheduled or can be loaded only once. There are options to create further jobs or to export the job back to the back-end planning system.

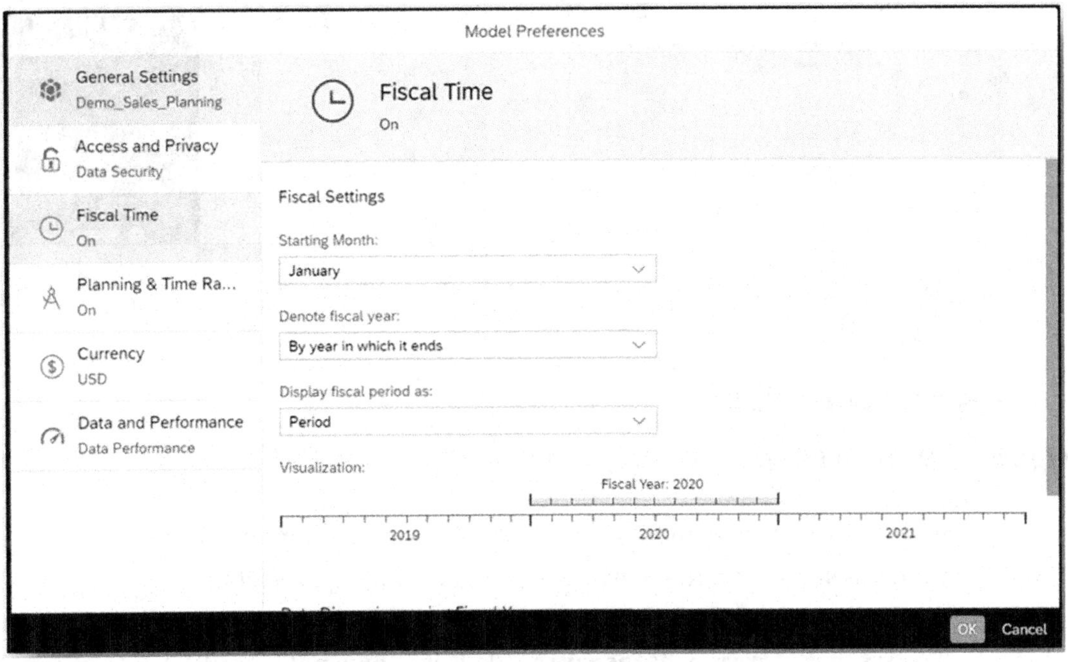

Figure 4-19. *Planning Model Preferences*

Further options need to be set to ensure that the planning model functions as expected. Click on the **Model Preferences** from the main model menu as shown in Figure 4-19. We have learned about Model Preferences in detail in Chapter 3 under the section **"Model Preferences."**

Since this is a planning enabled model, the preferences would hold information on time as well as planning. Set the options for the fiscal period as well as the planning and time options. This will ensure that the planning information is displayed as expected while we work on a planning story or a value driver tree, which we will learn in the coming sections.

Add Forecast Data to the Model

Plans need to have forecast data in addition to actual data. If actuals represent the current situation of where the organization stands, forecast data represents how the organization has planned for future periods. Again, this data can be directly acquired from back-end systems like S/4HANA embedded planning or SAP BPC or can be uploaded from a file. For demonstration purposes, we will use data imported from a file from the local file system. The entire process is shown in Figure 4-20.

Figure 4-20. *Map Forecast Data Elements*

Step 1: From the data management tab, click on the "**Import**" data option as shown. From the two options that come up, select the option to "**Import data from a file.**" The other option is to "**Import data from a datasource.**" This is shown in Figure 4-20.

Step 2: The File Import window comes up. Select the appropriate file name **"Sales_data_sample_planning.xlsx"** and the Forecast tab. Click "**Open.**"

Step 3: Select the tab from which to "**Import Data**". In this particular case, we will select data from the forecast tab. Click "**Import.**"

Step 4: The draft data source is created as shown in Figure 4-20.

Step 5: However, the data is not yet mapped to the correct fields in the mode. To do so, double-click on the **"Draft Data Source"** and the mapping window comes up. Ensure that the objects on the right-hand side are appropriately mapped to the fields on the left. Ensure that each of the field formats are also correctly set.

Once the mapping and format is correctly set, validate the data to iron out any discrepancies in the data and click on **"Finish Mapping."** The new job is created and run. The Forecast data is now loaded into the Model.

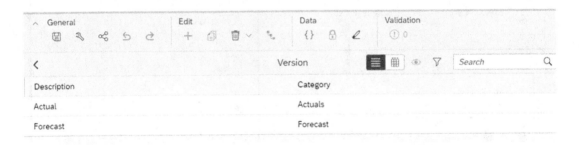

Figure 4-21. *Planning Model Categories*

Checking the version now shows two categories for which data has been loaded, Actuals and Forecast. This is shown in Figure 4-21.

We have seen how a planning model can be created and data loaded appropriately. Now this data needs to be explored appropriately within a story. Though there are multiple ways of exploring planning data, including gathering the data in a grid page, we will learn to present the data as a story to demonstrate the possibilities of analyzing planning data with analytical data.

In the next section, let us learn the step-by-step process to create a story on the Model created to explore the planning data.

Create Story on Planning Model

We have learned how to create a planning model in the previous section. A story on a planning model is similar to an analytical story. The process for creating a story on a planning model is shown in Figure 4-22.

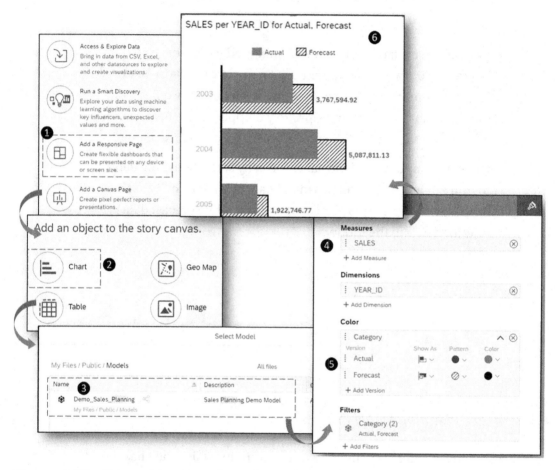

Figure 4-22. *Create Story Based on Planning Model*

The story can be built with any of the options including Canvas, Responsive, and Grid. As we have seen in previous sections "Building Interactive Stories", let us learn to create the Story with a responsive page. This story can be explored with equal dexterity in a mobile application as well. This is shown in Figure 4-22.

Step 1: Add a Responsive Page.

Step 2: Select the "**Chart**" option to input the data.

Step 3: Select the **"Model."** We have loaded data into the planning model **"Demo_Sales_Planning"**.

Step 4: From the Story Builder, select the measure and the dimension. We have selected **"Sales"** and **"Year_ID".**

Step 5: From the color selection option, select the category to represent the colors. Select the versions to represent the colors. In this particular case, we have selected "**Actual**" and **"Forecast"**.

Step 6: Click on "Save" to save the story. The variances are as shown in the Figure 4-22.

Variances help in checking divergences from the baseline or expected values. By adding a variance to a chart, the differences from the base values can be quickly brought forth.

Adding Variance to Planning Story

Let's start the process as shown in Figure 4-23.

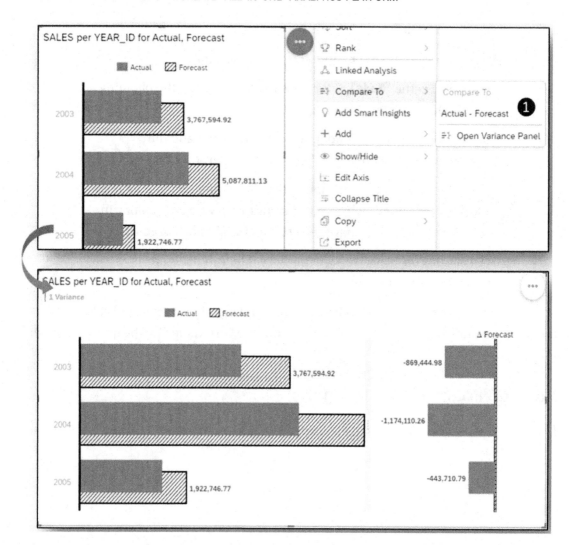

Figure 4-23. *Add variance to planning story*

Step 1: Click on the three dots to bring up the context menu for
the graph. Click on the **"Compare To"** option. Further select the
"Actual - Forecast" option. The variances are displayed as
shown in the reg bar. This can be interpreted as the difference
between the forecasted value to the actual value, as shown in
Figure 4-23.

SAC enables all Enterprise-wide collaborative planning including core planning
features like spreading, distribution, and allocation. Spreading is the process of assigning
funds at a higher level and then assigning them to a lower level: for example, moving

funds from the year level to months. Distribution is the process of assigning funds to other cells at the same level in the hierarchy. Allocation is the process of assigning funds based on certain business rules. All of the above and more are dedicated planning functionalities and outside the scope of this book. We will focus on how Planning capabilities of SAC enable organizations (ABC Inc.) to bring all analytics capabilities under the purview of a single platform enabling tighter governance and rapid decision-making.

Providing a single platform for operational, strategic, and financial planning, SAC enables enterprises to improve alignment to the plan as well as performance. SAC Planning also enables integration with enterprise applications and brings together all lines of business and enterprise functions including Sales, HR, Finance, and IT. This is achieved with live connectivity to enterprise applications we have seen in detail in Chapter 3 under the section "Connections." Enabling simulations through the Value Driver Tree, enterprise planners are able to visualize the impact of changes to goals set in the plans. We will learn about Value Driver Trees in detail in the next section.

We have learned how to create a planning model and explore data by creating a story and variances on it.

With the above, we have now seen two aspects of "All-in-One" Analytics platforms by learning how Business Intelligence and Planning can be integrated over a single platform.

Let us now learn how to simulate data and create what-if scenarios with Value Driver Trees.

Value driver trees enable what-if analysis within a particular dataset. This tool is especially helpful in terms of planning data, which can be used for simulating the effect of a particular value across a time span. A value driver tree works on the concept of nodes and branches. Nodes can be combined with branches to see the effect of changes to summarizing or union nodes.

As shown in Figure 4-24, the **"Total Sales"** is a combination of Motorcycle Sales and Truck Sales. Let us learn how the impact of change in Motorcycle Sales has an impact on the Total Sales and how this change can act as the base for future planning on sales strategy and subsequent goals. Let us now explore the step-by-step process to create a Value Driver Tree.

Creating Value Driver Tree

Let's follow the process as shown in Figure 4-24.

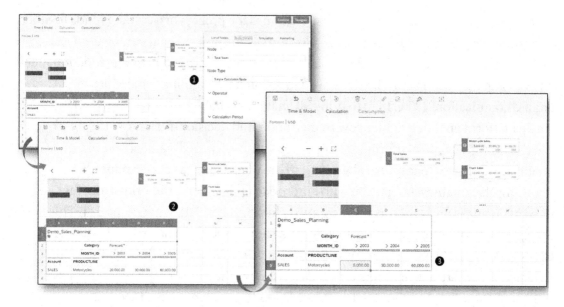

Figure 4-24. *Create Value Driver Tree*

Step 1: Click on the main menu and click create **"Value Driver Tree."** Create the nodes as shown in the first step. Create the tree for calculation as shown in the first step in Figure 4-24. In this particular example, we have created a formula for the node to add the values for sales of motorcycles and trucks.

Step 2: This can then be copied into the **Consumption Tab**.

Step 3: The information put in this tab can be changed to see the impact on the final node. All the what-if changes can be made to the child nodes, and the impact of the simulation can be seen in the overall structure.

The above is just an example and enterprise structures would be far more complex involving multiple nodes and calculations.

In the section above, we learned how to simulate data with Value Driver Trees. Let us now learn the step-by-step process to create the last component for "All-in-One" Analytics Platform, which is the Digital Boardroom.

Creating a Digital Boardroom

The Digital Boardroom is a dynamic presentation tool for executives to explore data in real time and build contextual insights. Built specifically to deliver LOB data from S/4HANA and other systems, the analytics are delivered over high-definition touch screens directly into the boardroom. Executives can use the data to arrive at data-driven decisions in real time without having to wait for answers to come from the underlying LOBs. Moreover, the Digital Boardroom is device agnostic and can be viewed equally easily over a mobile device.

Note   The Digital Boardroom feature is available with an additional license.

Digital Boardrooms are of two types, both of which follow nearly the same process to create:

- Digital Boardroom Agenda:

  Digital Boardroom allows for the creation of an end-to-end agenda with who would be presenting which topic and at what time. This agenda can be shared over collaboration, and the entire boardroom discussion structure can be controlled through the Digital Boardroom tool.

- Digital Boardroom Presentation: Dashboard

  The Presentation is more like a Dashboard, which can be presented by one or more people in the boardroom.

For the purpose of demonstration, we will learn how a Digital Boardroom Agenda can be created. Here is the step-by-step process to create a Digital Boardroom Agenda:

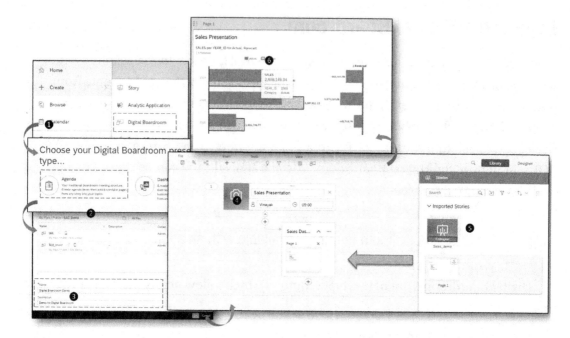

Figure 4-25. *Create Digital Boardroom Agenda*

Step 1: From the main menu, click on **CREATE** and select "**Digital Boardroom**" as shown in Figure 4-25.

Step 2: Select the option to "**Create an agenda.**"

Step 3: Give a name to the Digital Boardroom Agenda as shown in Figure 4-25.

Step 4: The agenda screen comes up. The Digital Boardroom is essentially composed of multiple components of existing stories. In the **Agenda** page, set up the agenda as shown. Select the name of the presentation, insert a picture if needed, and set up the branch.

Step 5: In the subsequent agenda, click on the box to add pages from existing stories.

Drag and drop the story into the box to set up the presentation.

Step 6: Save the boardroom agenda and click on **view** to view the Digital Boardroom Agenda. The agenda can be accessed from the file manager and presented directly onto the touch screen in the boardroom.

We have seen in the above sections how stories can be created. All objects including stories, models, digital boardroom agendas and presentations, and analytic applications can be explored via the File Browser in SAC. The File Browser is as shown in Figure 4-26.

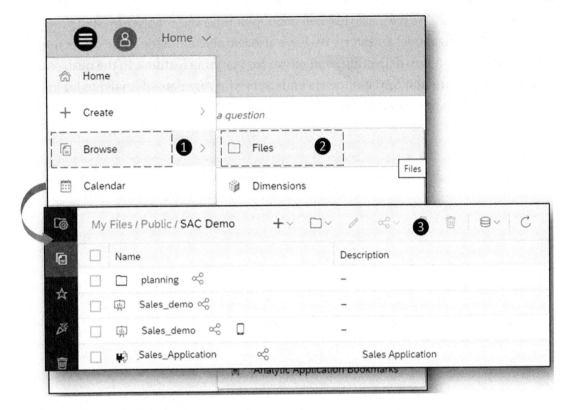

Figure 4-26. *File Browser*

Objects can be explored from the Main Menu by clicking Browse as shown in Figure 4-26 under ❶. Clicking on Files as shown in ❷ takes us to the file structure as shown in ❸. The files can be clicked to open.

We have now learned all the four components to bring together an integrated "All-in-One" Analytics platform. viz., Business Intelligence delivers deep insights, planning for enterprise planning and analytics, value driver tree and digital boardroom for data analysis, and presentation capabilities for the platform.

Summary

In this chapter, we have learned how SAC provides a single platform for all analytics. We have learned in depth the step-by-step process to create stories and digital boardroom agendas. We have also learned about planning and the step-by-step process to create a planning model and story. We have also learned about the value driver tree and how it contributes to delivering simulations for planning features. In the next, chapter we will learn how SAC empowers end users with Augmented Analytics for in-depth data analysis and storytelling.

CHAPTER 5

Exploit "Augmented Analytics" Capability of SAC

The analytics solutions at ABC Inc. are not in the same league with analytics solutions used by the competition of ABC Inc.: for example, the Use of a Mobile device, Use of Machine Learning, or Predictive algorithms in Analytical applications. Further, the solutions are not even meeting operational expectations, let alone enabling a competitive edge from the vast pool of information. Most of the Advanced Analytics applications are a black box to most of the business users and even the IT team, as only data scientists and a few experts understand them who are engaged on an ad hoc basis, making these solutions difficult to be used and remain accessible to limited people with advanced technical knowledge. One of the other occupying challenges for ABC Inc. is slow database performance leading to a number of internal challenges, including business intelligence security issues, fragmented processes, poor interdepartmental communication, and severe reporting lags. ABC Inc. is keen that their IT costs are part of the ongoing operational expenses by paying a subscription cost rather than one massive, big capital expense that requires approval of multiple stakeholders, including board members. There are many additional challenges that ABC Inc. deals with: for instance, Business Planning is done in MS Excel, and its iteration takes a huge toll on people involved due to a lack of collaboration.

Over the years, advanced analytics has emerged as the cornerstone of driving business competitiveness. Analytics tools not only offer a self-service platform but are also expected to deliver a platform that offers advanced analytics and can be scaled rapidly.

© Vinayak Gole, Shreekant Shiralkar 2020
V. Gole and S. Shiralkar, *Empower Decision Makers with SAP Analytics Cloud*,
https://doi.org/10.1007/978-1-4842-6097-5_5

Data analysis is no longer a black box, and normal business users with analytical decision support, aka citizen data scientists, have been observed delivering multiple business insights with available data. For example, an end user who works with business logistics day in and day out, when armed with data analysis capabilities, can deliver an optimal path to material movement and storage. Trend analysis will enable sales executives to tailor their sales pitches for each customer. Customer analysis will allow marketing executives to target the right audience with the right product and build brand loyalty. With the base of data analytics, businesses can deliver value to customers, both internally and externally. But the prerequisite for such a scenario would be having the right technology. Precious time is spent by a business putting up requirements for the IT team, who then reaches out to data scientists, and then comes up with a dashboard. The intended platform should encourage real-time collaboration between teams and allow end users to take control of data to drive true data-driven decisions with minimal IT intervention.

One of the vision items for ABC Inc. is to have a modern analytics platform to encourage citizen data scientists within the organization. However, this would involve a steep learning curve, and hence the task is to have a platform that would enable augmented analytics for end users. Augmented analytics is a feature of modern analytics, which enables the use of machine learning technologies to enable end users to navigate and gain from advanced analytics most effortlessly. Algorithms within machine learning technologies are typically black boxed to insulate the end user from the complexities of these technologies. Some of the typical use cases for augmented analytics are for predictive forecasting, data-driven decision-making, and advanced storytelling.

Augmented Analytics assists end users with machine learning and predictive technologies for developing trends analyses as well as insights. The platform should provide the right encouragement and direction for end users to benefit the most by exploring the data. Automation and machine learning technologies should be inclusive and not just limited to designated data scientists. Data and trends should be available for everyone to analyze not just on desktops but also within boardrooms where executives would not have to spend valuable time waiting for data to be discovered, but for questions to be answered in real time. The data analysis should support easy navigation and also handhold the executives as they do a 360-degree traversal through the data.

And all of the above features should be available in a cost-effective solution, preferably in a scalable and flexible, pay-as-you-go solution.

With this vision, ABC Inc. is interested in an analytics platform that meets these requirements:

1. **Has advanced augmented analytics capabilities:**

   Advanced analytics offer the edge against competition in the modern world. A data warehouse and classic business intelligence do not typically suffice in today's world. Advanced analytics including predictive and automation technologies are the norm for driving business in a world driven by technology.

2. **Is scalable:**

   The new system should be able to scale rapidly with minimum disruption to the existing landscape. It should be flexible and rapidly upgradable with the latest technologies with low investment and without any of the issues typically encountered during an upgrade.

3. **Collaborative:**

   The landscape should enable inline collaboration instead of relying on external systems for interaction between users.

4. **Cost effective:**

   It should be low on the budget and not exceed budget expectations.

Alignment to Specific SAP Analytics Cloud Capability

As we have read in the previous section, ABC Inc. is facing a major challenge in terms of having an outdated system, lacking modern analytical capabilities. As the world wakes up to the power of machine learning algorithms using statistical models to predict future trends, ABC Inc. is at a major disadvantage compared to its peers.

SAC has been built as a cloud-native application over the SAP Cloud platform and utilizes machine learning technology intrinsic to the underlying platform for delivering effortless use and experience to the end user. The augmented analytics capabilities available within SAC use techniques such as data mining, statistical modeling, and machine learning to present the end user with a forecasted value based on historical data. SAC's augmented analytics capabilities will thus help ABC Inc. enable a modern analytics landscape that empowers end users with "Insights to Action." The Augmented Analytics capabilities of SAC are collectively known as "Smart Assist."

Let us now learn about some of the advantages of implementing SAC for ABC Inc.

1. **Faster insights to action:**

   One of the fallacies of the current landscape for analytics is the workflow-based change management and dependency of the business on IT for its reporting needs. With minimal self-service available, end users have to depend highly on IT services for building new reports and implementing changes to the reports, which go through a lengthy approval and deployment process. Many end users have resorted to creating their own datasets in spreadsheets to work better with data on their local computers. This has led to a non-standardized approach to data analysis with considerable time spent on reconciliation.

   The augmented analytics of SAC enable end users to not only build stories using machine language technologies but also converse with the analytics landscape in natural language. The ease of exploring data brought about by Smart Discovery and Search to Insight features enables end users to focus on specific data points and explore data contextually with domain knowledge. Predictive Forecast and Smart Insights allow end users to rapidly convert analytics into actionable insights by creating a plan with a focus on forecasted values. And all of the above is in a secure landscape accessible from anywhere with minimal intervention from IT.

   Considerable time can be reduced in reconciliation and follow-ups, and end users can spend time saved on high-value tasks, delivering better value to business goals.

2. **Low learning curve:**

   In the current scenario, ABC Inc. has multiple teams with specific functions. End users have always been reluctant to take up advanced analytics due to the steep learning curve in new age technologies.

SAC enables the use of modern advanced analytics, the underlying technological complications of which are shielded from end users. They can derive all the advantages without having to delve deeper into understanding machine learning and statistical models. Augmented analytics enables end users to include forecasts and simulations directly into their stories. Stories can be presented to management thought the single SAC platform.

With the low learning curve, end users can save time, avoid redundancy, and make the best use of available data and also make data-driven decisions.

3. **Reduced redundancy:**

In the current scenario, there are multiple teams who work on the same KPIs with no collaboration. Moreover, due to the legacy system, there is rigidity in exploring specific data, and end users have to rely on holistic reports to extract data for their analytical needs. Some of these reports are scheduled, which are seldom used by end users. Executive dashboards and simulations are built in Office applications, allowing minimal flexibility and multiple reviews before a version can be finalized and presented.

SAC's Smart Discovery and Smart Insight features, when combined with the powerful storytelling capability, enable end users to build focused reports and presentations. Executive dashboards can be built directly over digital boardrooms, which allow for real-time data exploration. Deeper insights are presented by Smart Insights.

End users can now rely on Smart features to simulate business outcomes and strategize accordingly. Multiple teams can collaborate over a platform and search for insight to enable end users to deliver answers to business questions with minimal dependency on IT.

The primary components of Smart Assist include the following:

1. Search to Insight

2. Smart Insights

3. Predictive Forecasting

4. Smart Discovery

5. Smart Predict

6. Smart Grouping

Now we will learn, in detail, about SAC's "Smart Assist."

1. **Search to Insight:**

In the age of Data, on one hand, Enterprises are overwhelmed with humongous quantities of data. Data analysis, on the other hand, is riddled with questions but no clear definition of answers. A tool that would allow business analysts to search through data and mix and mash analyses across datasets would be of the utmost use in such a scenario.

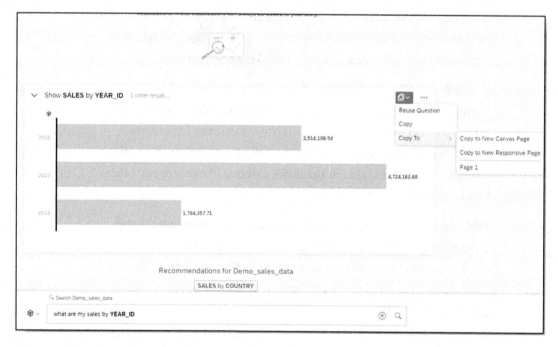

Figure 5-1. *Search to Insight*

Search to Insight is one of such capabilities of SAC that allows end users to get quick answers to business questions. It also allows end users to converse in Natural Language, completely shielding end users from the intricacies of database querying and data mashing. Natural language along with autocomplete words and phrases also allows end users to come up with contextual stories for their areas of investigation, allowing for quicker turnaround times and faster business decisions. A typical Search to Insight window is shown in Figure 5-1. This shows how a question can be asked in a Natural Language, which is then answered by a Search to Insight.

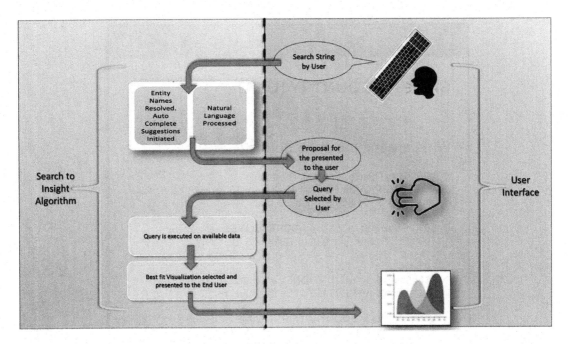

Figure 5-2. *Search to Insight – High-Level Flow*

The high-level flow for the Search to Insight is shown in Figure 5-2. The process starts by a user entering a search string. This string is resolved against indexes available and autocomplete suggestions are provided. Once the user completes the search string, the Natural Language is processed and further proposals are offered to the user for selecting the right query. Once the user selects the query, it is executed against the available data and displayed with the best fit visualization type. This is sent back to the user window where the user can either finalize or delve deeper with more questions. The process then becomes iterative.

2. **Smart Insights:**

Smart Insights feature uses machine learning to automatically provide insights on available data. Let us consider an example where the end user has already built a story or is referring to a pre-built dashboard. For every additional query that the end user has, the traditional route would be to get the requirements to the IT team for building into the dashboard. In some cases, the query might be for one-time use only and might not require significant time to be spent on building the same. Alternatively, the end user might want to delve deeper into the data but does not have a well-defined starting point.

Figure 5-3. *Smart Insights*

A typical Smart Insights window is shown in Figure 5-3. Smart Insights enable instant explanations for the end user by bringing forth top contributors to any measure as shown in this figure. These insights are available at the click of a button without the end user having to spend valuable time digging through data, slicing and dicing it, allowing the end user to spend time on deliverables and tasks of higher value.

3. **Predictive Forecasting:**

Predictive technologies have traditionally been the forte of data scientists, requiring very specialized skills and training. Custom-built statistical models would be built to identify trends and outliers, which would eventually be used for forecasting and influencing business decisions. SAC provides built-in ready-to-use forecasting tools for predictive forecasting that can be enabled from the chart options. Supported charting options are Time Series Charts, Line Charts, and Planning Grids.

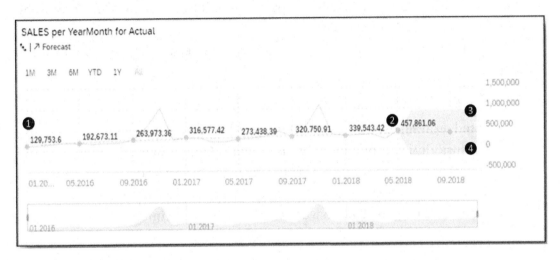

Figure 5-4. *Predictive Forecast*

Figure 5-4 shows a typical Predictive Forecast Window, which shows the linear forecast of Sales based on Year and Month. Predictive forecasting is available in three forecasting methods.

a. **Automatic Forecast:**

SAP's internal algorithm determines the forecast for the period: algorithm analysis trends, cycles, and data to come up with the best fit forecast. As a simple equation:

Automatic Forecast = Trend + Cycles + Fluctuation + Residual.

This algorithm ensures the best forecast for available historical data.

b. **Linear Regression:**

Linear regression follows a more linear trend and enables the forecast to be more linear and simpler in nature. In simple terms, it continues to expand the current trend with a steady error margin.

c. **Triple Exponential Smoothing:**

Exponential Smoothing allows smoothing of time series data into patterns. Triple exponential smoothing adds a variable for cycles and allows better results for the forecasted variable.

The predictive forecast function also helps to determine the quality of the forecast by rating the forecast over a 5-point scale.

As shown in Figure 5-4, the Predive Forecast can be understood by considering four components:

i. **Historical Data**

Based on available data, the chart is plotted across the time axis. The historical data plotting starts as shown in ❶ and will continue until ❷ as shown in Figure 5-4. The trend of the historical data defines the plot that will be followed for the Predictive Forecast. As shown in this figure, **Sales** have been plotted against **Time**.

ii. **Predictive Forecast:**

Predictive forecasts for **Sales** start where the historical data plot ends. From Figure 5-4, the predictive forecast starts from ❷ and continues in a straight line.

iii. **Upper Confidence Level:**

The Upper confidence level determines the upper limit for the level of errors. This is shown in ❸ in the figure.

iv. **Lower Confidence Level:**

The Lower confidence level determines the lower limit for the level of errors. This is shown under ❹ in the figure.

The forecasted value should ideally be between these two levels.

4.   **Smart Discovery:**

Smart Discovery allows end users to either create a story detailing key influencers, anomalies, and simulation options or add all the above to an already existing story. Smart Discovery allows end users to rapidly build simulations for critical business decisions without being exposed to underlying advanced technologies.

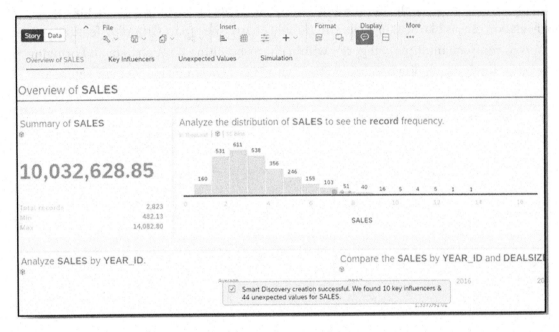

Figure 5-5. *Smart Discovery*

Figure 5-5 shows a typical Smart Discovery Overview Page, which has the summary of sales along with Smart Insights into the variable. In addition to the Overview page, there are other features in terms of pages, which we shall explore in the section on the step-by-step process to create a Smart Discovery.

Smart Discovery can be run from either within a Story or from the Story Builder Wizard. Components of Smart Discovery can thus be invoked and included within an existing story or a new Smart Discovery Story created independently.

5.   **Smart Predict:**

SAC now features a fully functional predictive module that enables end users to build actionable predictions with ready-to-use statistical models. Data analysts can train these predictive models to come up with forecasted values, trends, and anomalies allowing businesses to align their strategies. Delivering true data-driven decisions, this is one of the most powerful features of SAC when combined with other analytical features. We shall see further details of Smart Predict in Chapter 7 where the entire chapter is dedicated to predictive technologies in SAC.

6. **Smart Grouping**

Smart Grouping is another feature of the Smart Assist portfolio, which automatically enables end users to create groups and categories. Especially useful while using co-relation charts like Scatter plots and Bubble charts, the Smart Grouping feature assists users to compare multiple categories within charts by sifting through data and bringing together similar categories. A cluster chart with Smart Grouping is shown in Figure 5-6.

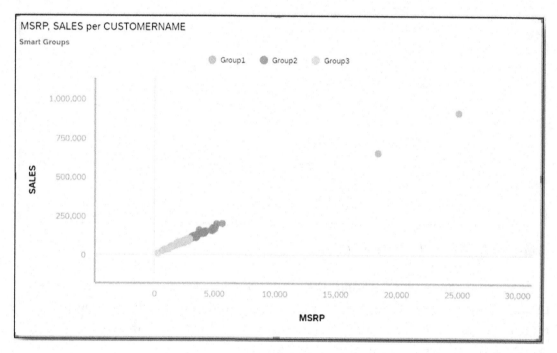

Figure 5-6. *Smart Grouping*

Implementing SAC will thus allow ABC Inc. to move up the advanced analytics value chain with minimal disruption for end users and analysts. Advanced analytics features delivered by SAC will enable the organization to move toward being a data-centric organization.

In the next section, we will learn the step-by-step procedure to build each of these features except Smart Predict, which we will learn about in the coming chapters.

As we discussed in the last section, augmented analytics is one of the key tenets for the modern Analytics Landscape envisioned by the customer. We learned how augmented analytics can enable end users with the ability to quickly develop data to insight and deliver rapid decision-making capability. Let us now learn the step-by-step process for creating each of the augmented analytics capabilities.

Creating Augmented Analytics

1. **Creating Search to Insight**

One of the frontline features of augmented analytics in SAC is the Search to Insight. Search to insight can be used to explore data independently or within an already existing story. Let us first learn about the Search to Insight Application.

The Search to Insight Window is shown in Figure 5-7.

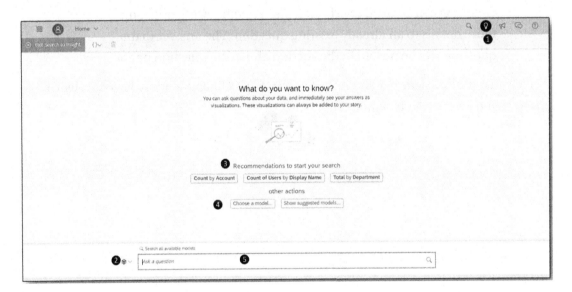

Figure 5-7. *Search to Insight Window*

1. The bulb icon ❶ is from where the application can be initiated. The option is available in the main menu and can be accessed from the Story Builder.

2. **Model Selector:**

   If Search to Insight is used for data analysis without a corresponding model, the model selector can be used to select the appropriate model on which to perform analysis as shown in ❷.

3. **Recommendations:**

   Option ❸ uses smart technologies to provide the end user with a context for the search. The most frequently used searches come up first, followed by other actions.

4. **Other actions:**

   Option ❹ enables selecting a model as well as viewing the model most suitable for the search suggestions from the previous section.

5. **Question panel:**

   Option ❺ can be used to ask a question to the system or to add more details to an already existing question. The answer to the question will be posted in the section above the question panel.

Now that we have learned about the Search to Insight application, let's now learn how to create a Search to Insight in Figure 5-8.

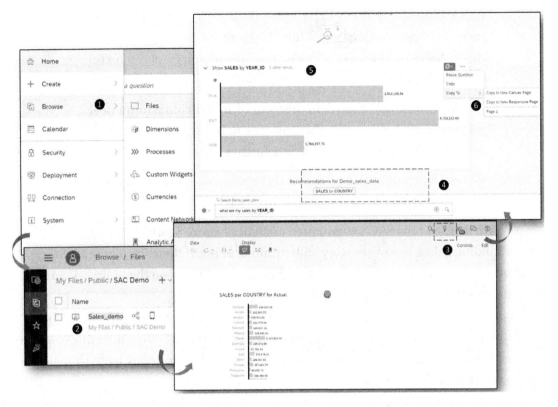

Figure 5-8. *Create Search to Insight*

Let us now learn how to create Search to Insight to add additional information to an already existing story.

> Step 1: Click on **"Browse"** and then **"Files"** from the Main Menu as shown in Figure 5-8.

> Step 2: Open the Story "Sales_Demo". We will continue to leverage the Story we already created in Chapter 4 under the section **"Step-by-Step Process to Build Interactive Stories."** Click on **"Edit"** to bring up the **"Story Builder"** Window, navigate to the top menu, and click on the bulb icon. This brings up the **Search to Insight** window.

> Step 3: The system puts forth the best recommendations for the query, which can be accessed from the recommendations bar as shown in the Figure 5-8.

137

Step 4: If the query needs to be customized, click on the question panel and type the query in natural language: for example, "What are my **Sales** by **Year_id**?" Click on the magnifying glass ⌕ icon to start the query as shown in the Figure 5-8.

Step 5: The relevant answer in the form of an individual numerical point or graph is put forth. This answer can be further refined by reframing the question or by adding further dimensions.

Step 6: The chart can also be directly copied into a story or a page by copying from the menu as show in Figure 5-8.

We have learned how SAC's augmented analytics feature Search to Insight can enable building a data discovery story with minimal effort. If there is no existing Story, the above process can be followed to query the data by clicking on the bulb 💡 icon from the main menu as shown in Figure 5-8.

Smart Insights enables end users to quickly delve deeper into available data. This augmented analytics feature uses an in-built algorithm to derive more meaning from the data in the story. Let us now learn the step-by-step procedure to create a Smart Insights feature within an existing story.

2. **Creating Smart Insights:**

Smart Insights provides additional information from the existing dataset. Enabling end users to better utilize the self-service capabilities of SAC, Smart Insights uses an internal algorithm to sift through the data and deliver additional insights for the end user to choose from. The user then has the liberty to either choose from the provided insights, explore them further, include them in the story, or ignore the insights provided based on relevance.

Let us now learn the step-by-step process to create Smart Insights in SAC as shown in Figure 5-9.

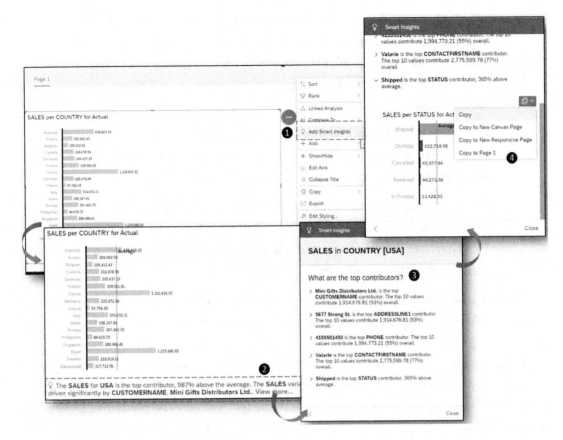

Figure 5-9. *Add Smart Insights*

Step 1: Browse to the existing story or create a new one. We will continue to leverage the Story we already created in Chapter 4 under the section **"Step-by-Step Process to Build Interactive Stories,"** which is **"Sales_Demo"** by clicking on the **"Browse"** and **"Files"** and navigating to the Story. Click on **"Edit"** to bring up the **"Story Builder"** Window. Select an object (graph or chart) from the story and click on the **Chart Menu** with the three dots as shown in Figure 5-9.

Step 2: This **Chart Menu** brings up other options for the selected chart. Select the option **"Add Smart Insights"** as shown in the Figure 5-9.

Step 3: The underlying algorithm now analyzes the background data and the contextual information available and brings up **Smart Insights** as shown in Figure 5-9. In this particular example, the **Smart Insights** on sales come up with the option to **"View More"** as shown in the figure.

The entire set of Smart Insights is displayed. In this particular example, the context is the top contributor. Hence the algorithm brings up all the options for top contributors within the dataset.

Step 4: Expanding a particular section, for example, **"Shipped"** will bring in further details, which can be in the form of a chart. This chart can directly be pasted into the existing or new story or a dashboard for further contextual analysis.

We have now learned the step-by-step process of creating Smart Insights, which is an Augmented Analytics feature of SAC, and how it helps end users get further insights into data.

As discussed in the section "Alignment to Specific SAP Analytics Cloud Capability," one of the key tenets of modern predictive analytics is to allow end users to add visualize forecasts without the need to interpret underlying technology. The predictive forecasting algorithm allows the end user to analyze past data to predict future trends.

The Predictive Forecast feature of SAC allows end users to quickly analyze future trends based on historical data forecasting. Let us learn how SAC Predictive Forecasting is able to help end users understand business KPI forecasts.

3. **Creating Predictive Forecast:**

Predictive Forecast enables end users to include forecasting capabilities within the trend seen in the data. Built into the forecasting capabilities, it is easier to implement and can be included without having to build an end-to-end predictive scenario. The Predictive Forecasting capability is extremely essential for end users who want to make a rapid decision on the available data and immediate repercussions of any changes to the parameters.

Let us now learn the step-by-step process to create a predictive forecast within a chart in a story as shown in Figure 5-10.

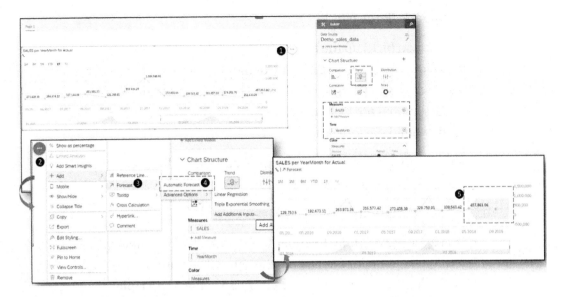

Figure 5-10. *Predictive Forecast*

Step 1: Browse to the existing story or create a new one. We will continue to leverage the Story we already created in Chapter 4 under the section **"Step-by-Step Process to Build Interactive Stories,"** which is **"Sales_Demo"** by clicking on **"Browse"** and **"Files"** and navigating to the Story. Click on **"Edit"** to bring up the **"Story Builder"** Window. Select an object (graph or chart) from the story and click on the **Chart Menu.** Add a Time Series chart to the **Story** with **Sales** as a measure and **YearMonth** as a Time dimension as shown in Figure 5-10.

Step 2: Click the three-dot menu of the chart to bring up the **Chart Menu.** This is as shown in the figure.

Step 3: Click on **Add,** which brings up further options for adding features to the existing chart. Click on **Forecast**. The **Forecast** options come up. There are two options for the forecast.

a. **Automatic Forecast:**

This option picks up the best option available for forecasting the value of the measure in consideration. This option is selected by default, and end users need not spend much time understanding the technology behind the same of the logic used to build the forecast.

The underlying time series algorithm automatically selects the best fitting time series model.

b. **Advanced options:**

This option gives more freedom to the end user to manually select the option for forecasting the value. The options available for forecasting are **Linear Regression** and **Triple Exponential Smoothening**. There is another option for advanced options that can be used to tweak the parameters further.

Step 4: *(Automatic)* Let us select the **Automatic Forecast** for the purpose of demonstrating the capabilities of Predictive Forecast. This is shown in Figure 5-10.

Step 5: *(Automatic)* The Forecast comes up as shown in Figure 5-10. The straight line depicts the most likely path for sales growth with a margin of error on both sides between which the predictive forecasting might vary.

Step 6: *(Automatic)* The quality of the forecast can be checked by clicking on the **Forecast** option under the chart header. Quality of forecast is predicted as a score out of 5. As can be seen from Figure 5-11, the quality score of the forecast is 4 out of 5, which is a good score and hence the forecast can be considered of good quality and reliable.

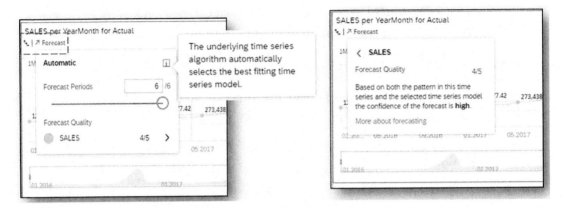

Figure 5-11. *(Automatic): Predictive Forecast quality*

Step 4: *(**Advanced**)* Let's select the **Automatic Forecast** for the purpose of demonstrating the capabilities of Predictive Forecast. This is as shown in Figure 5-12.

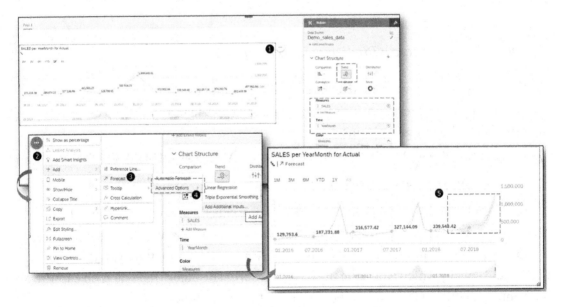

Figure 5-12. *(Advanced): Predictive Forecast*

Step 5: *(**Advanced**)* The Forecast comes up as shown in Figure 5-12 *(**advanced**)*. The straight line depicts the most likely path for sales growth with a margin of error on both sides between which the predictive forecasting might vary.

Step 6: *(Advanced)* The quality of the forecast can be checked by clicking on the **Forecast** option under the chart header. Quality of forecast is predicted as a score out of 5. As can be seen from Figure 5-13 *(advanced)*, the quality score of the forecast is 5 out of 5, which is a good score and hence the forecast can be considered of good quality and reliable.

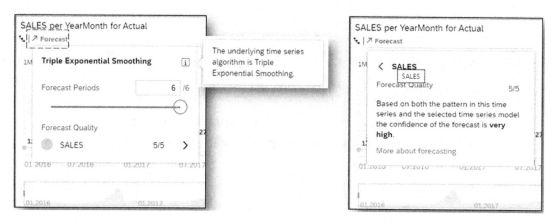

Figure 5-13. *(Advanced): Predictive Forecast quality*

We have learned how Predictive Forecast enables end users to develop a future view for a measure in consideration. This augmented analytics feature will help end users make informed business decisions.

Let us now learn about the next Augmented Analytics feature in Smart Discovery.

4.   **Creating Smart Discovery:**

SAC's underlying algorithm for Smart Discovery sifts through the available data and creates a fully operational story complete with Simulations and deep insights. The complete process is automated and requires only minimal interaction with the end user who would specify which would be the base measure. The algorithm will then capture all the contextual information and build the Smart Discovery Story.

The Smart Discovery feature can be used to create a completely new independent story or can be also be used within an already created story to add further value to the story. Let us learn both of options of creating a Smart Discovery:

1. **Create a new Smart Discovery Story:**

This option to create a Smart Discovery Story enables end users to create a complete end to end analysis of a measure without any manual intervention. The created story provides information about the selected measure including outliers as well as a simulation page.

The entire Smart Discovery story creation process is as shown in Figure 5-14.

Figure 5-14. *Create Smart Discovery Story*

Step 1: Click on the main menu and click **"create"** and then click on **"Story."** The Story Window comes up. Select the option to **"Run a Smart Discovery."** We have seen the other options to create a story in Chapter 4 in the section "Step-by-Step Process to Build Interactive Stories."

Step 2: The **Smart Discovery** window comes up. This window has options to select the measure on which the smart discovery has to be built as shown in Figure 5-14.

Step 3: From the menu on the right in the **Discovery Settings** select measure **"Sales."** The Smart Discovery algorithm will build the complete story on this measure.

Step 4: Click on **"Run"** as shown in step 4. The algorithm now starts building the story based on the measure selected, which in this particular case is **"Sales."**

Step 5: Once the algorithm completes, the story is displayed as shown in the figure.

The created Smart Discovery Story has four tables, viz, Overview, Key Influencers, Unexpected Values, and Simulation. We will learn about each of these tables in further detail.

2. **Add Smart Discovery to an Existing Story:**

A Smart Discovery Process can also be initiated from within an already existing Story. This enables end users to add further information to their stories by bringing in all or individual pages from the Smart Discovery Story created. Consider an example where the end user already has a story on Sales. However, the user would like to add simulation to the story. The Smart Discovery can be initiated from within the existing story and only the Simulation Page can be retained while removing the other pages.

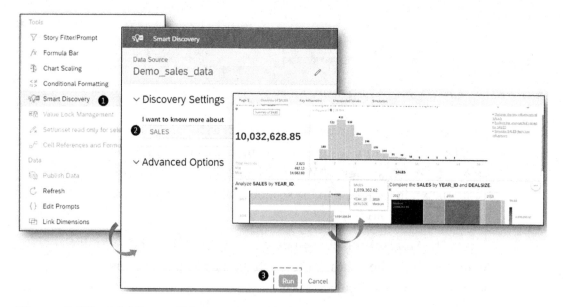

Figure 5-15. *Add Smart Discovery to existing story*

Step 1: For the purpose of demonstration, we will continue to use our existing **Story** "Sales_Demo". Click on **"Browse"** and **"Files"** from the main menu and then select the **Story**. Click on "Edit" to open the story in Edit Mode. Click on the top menu to bring up the **"Tools"/"More"** Option.

The **Tools** menu comes up with multiple options as shown in the figure. From this menu, select the **"Smart Discovery"** option as shown in Figure 5-15.

Step 2: The **Smart Discovery** window comes up. From the **Discovery Settings** options, select the measure on which the smart discovery has to be built. in this particular example we have selected **"Sales."**

Step 3: Click on **"Run"** as shown in Figure 5-15.

The Smart Discovery pages are added to the existing story as shown in Figure 5-15.

We have now learned how to create Smart Discovery Stories. Let us now learn the components of a typical Smart Discovery Story. The created Smart Discovery Story has four tables, viz, Overview, Key Influencers, Unexpected Values and Simulation.

Let us learn about each of these tabs in further detail.

3. **Components of Smart Discovery Story:**

Let us now learn about the Smart Discovery augmented analytics structure. Typically, a Smart Discovery story consider consists of four pages:

a. **Overview Page:**

This is the summary page and gives an overview of the measure selected. We have selected "Sales" as a measure and hence all the information about "Sales" shown with respect to the dimensions available in the model.

A typical Overview page for sales looks like the below in Figure 5-16.

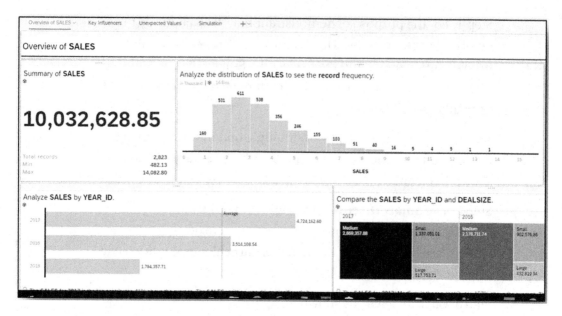

Figure 5-16. *Smart Discovery - Overview*

b. **Key Influencers:**

This page consists of the key factors that influence sales of this particular organization. All the KPI combinations are taken into consideration along with smart insights to add further value contextually. A typical Key Influencers page is shown in Figure 5-17, which explains what the key influencers for "Sales" measure are and how each of the key influencers impacts sales.

For example, how the "Dealsize" would have an impact on "Sales" can be seen in Figure 5-17.

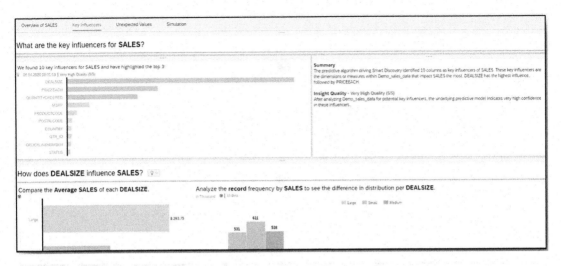

Figure 5-17. *Smart Discovery - Key Influencers*

c. **Unexpected Values:**

In any database, there are typically values that do not fall within the given criteria. These values are listed under the unexpected value category. These values need further investigation to examine if they are typical or just anomalies within the dataset.

A typical Unexpected Values page is shown in Figure 5-18, which shows a list of unexpected values ❶ and their impact on the current sales figures ❷.

This is shown in Figure 5-18.

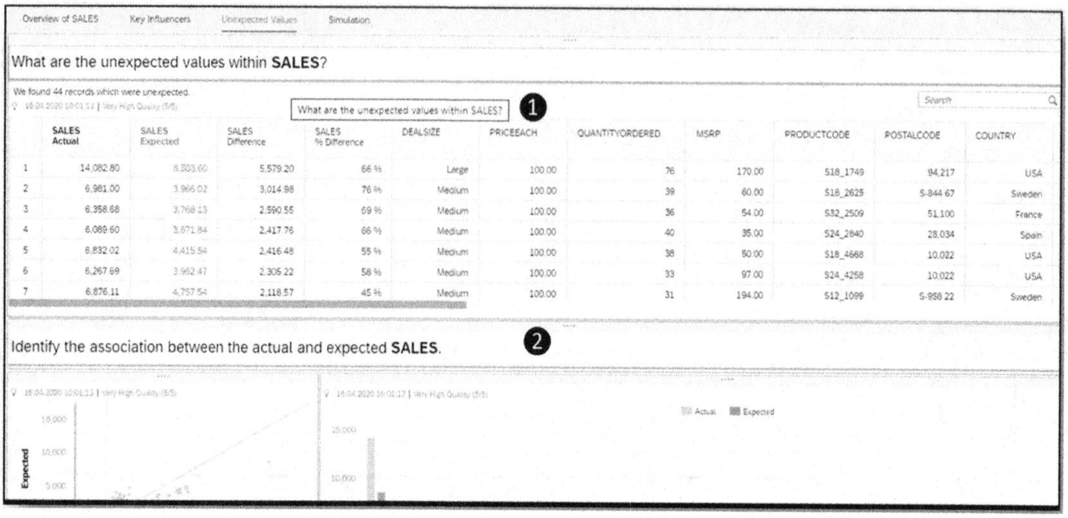

Figure 5-18 content:

Overview of SALES   Key Influencers   Unexpected Values   Simulation

What are the unexpected values within SALES?

We found 44 records which were unexpected.
18.04.2020 10:01:13 | Very High Quality (5/5)

What are the unexpected values within SALES? ❶

Search

	SALES Actual	SALES Expected	SALES Difference	SALES % Difference	DEALSIZE	PRICEEACH	QUANTITYORDERED	MSRP	PRODUCTCODE	POSTALCODE	COUNTRY
1	14,082.80	8,503.60	5,579.20	66 %	Large	100.00	76	170.00	S18_1749	94,217	USA
2	6,981.00	3,966.02	3,014.98	76 %	Medium	100.00	39	60.00	S18_2625	S-844 67	Sweden
3	6,358.68	3,768.13	2,590.55	69 %	Medium	100.00	36	54.00	S32_2509	51.100	France
4	6,089.60	3,671.84	2,417.76	66 %	Medium	100.00	40	35.00	S24_2840	28.034	Spain
5	6,832.02	4,415.54	2,416.48	55 %	Medium	100.00	38	80.00	S18_4668	10.022	USA
6	6,267.69	3,962.47	2,305.22	58 %	Medium	100.00	33	97.00	S24_4258	10.022	USA
7	6,876.11	4,757.54	2,118.57	45 %	Medium	100.00	31	194.00	S12_1099	S-958 22	Sweden

Identify the association between the actual and expected SALES. ❷

18.04.2020 10:01:13 | Very High Quality (5/5)   18.04.2020 10:01:13 | Very High Quality (5/5)   ▇ Actual ▇ Expected

Figure 5-18. *Smart Discovery - Unexpected Values*

d. **Simulation:**

The Smart Discovery augmented analytics option allows for
simulation by providing the end user with an interface to do
what-if analysis. The simulation allows for multiple values to be
changed to see the effect on the measure Sales.

A typical Simulation page is shown in Figure 5-19. The page shows
multiple influencers whose values can be changed to see the
impact on sales as shown in ❶ in Figure 5-19. The Simulation page
also shows which of the influencers have the maximum impact on
the values of sales.

This is shown in ❷ in the Figure 5-19.

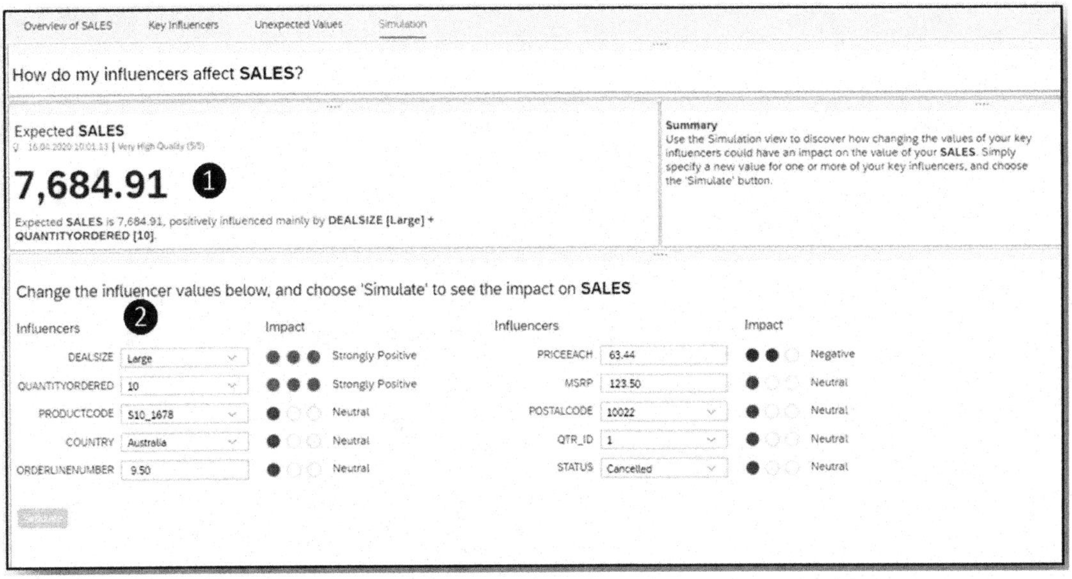

Figure 5-19. *Smart Discovery – Simulation*

Smart Discovery thus enables end users to be part of a modern analytics landscape that assists end users with data analysis and storytelling. Smart Discovery can be used by end users to rapidly build stories and analyze data trends.

Let us now learn about the next feature of Smart Assist, which is Smart Predict.

4.   **Implementing Smart Predict**

Smart Predict falls under the category of predictive analytics as well. It is an augmented analytics feature that allows the end user to work with predicted analytics without having in-depth knowledge of the underlying technology. We have an entire chapter dedicated to predictive technologies available within SAC.

We shall learn Smart Predict in depth in Chapter 7 where we will also learn about Predictive Technologies in SAC.

5.   **Create a Smart Grouping:**

Smart Grouping enables grouping together data into clusters that are similar. This enables end users to compare data across multiple groups while making critical decisions. A part of SAC's Smart Assist features, Smart Grouping enables providing the crucial visual aide while comparing multiple groups across the same set of measures.

Let s learn the step-by-step process to create a Smart Grouping in a Scatter Plot. The entire process is shown in Figure 5-20.

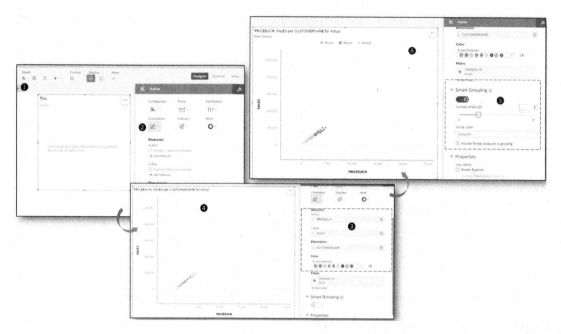

Figure 5-20. *Smart Grouping*

Step 1: Browse to the existing story or create a new one. We have seen the process to create or browse a story in Chapter 4. In this particular example, we have continued to use our **"Sales_Demo"** story. Click on **"Browse"** and **"Files"** and then select the **Story**.

Step 2: Insert a **Scatter Plot** in the **Story** as shown in Figure 5-20.

Step 3: A scatter plot chart would require two measures and a common dimension. Let us find out how the price of each item for customers impacted Sales. Add the measures **"Priceeach"** and **"Sales"** along with the dimension **"Customername"** as shown in Figure 5-20.

Step 4: The **Scatter Plot** chart comes up. However, all the data points are of the same color, leaving little differentiation.

Step 5: Let us now enable "**Smart Grouping**" from the Menu. Set up the number of groups by moving the slider. We have selected three groups within out dataset. This is as shown in the figure.

Step 6: The Scatter Plot now shows the data grouped on the basis of sets with common characteristics for Price of each item with respect to sales. The outliers can be clearly seen and presented in a separate color. This is as shown in Figure 5-20.

We have now learned about the Smart Assist or Augmented Analytics components within SAC. SAP has been rapidly ramping up the capabilities of the Augmented Analytics features of SAC by including support for Live Models. Though currently only selective features are supported for Smart Insight and Smart Discovery, Search to Insight is fully supported with Live Models. Smart Discovery is available only on Live HANA Models. Predictive forecasting can be made available for Live Models.

Summary

In this chapter, we have learned how SAC enables augmented analytics known as Smart Assist for enterprises. We have learned about the different options available for Smart Assist. We have also learned how to implement each of these options with a step-by-step process. In the next chapter, we will learn how SAC facilitates "Anytime Analytics," which enables data analytics over multiple devices round-the-clock ubiquitously.

CHAPTER 6

Develop SAC for an Anytime Available Platform

In Chapter 2 and specifically in item 5 of the section titled "Customer's Current Landscape and Pain Points," we discussed how legacy applications hosted on archaic and outdated infrastructure needed perpetual maintenance, causing lack of availability on a regular basis. ABC Inc.'s legacy analytics systems are primarily batch driven and do not provide access to real-time data. Further due to complexity of the landscape, any change or new analytical requirement takes considerable time to respond, and many end users therefore rely on using local solutions and working through their own on MS Excel-based options. Due to the lack of readily available data, these end users have to either defer the decision or rely on past experience, which is often not the best or most effective decision.

End users are also faced with similar issues with respect to having access to Business KPIs and dashboards at any point in time. Legacy landscapes are liable to multiple planned and unplanned downtimes due to regular updates and maintenance schedules. Senior executives are forced to rely on reports generated by schedules and do not have real-time access to data. This often is detrimental for crucial business decisions, which have to be made while considering the most recent data. In summary, the Analytical Solutions are plagued by an extremely sluggish response in meeting new requirements.

In a data-driven world, Business KPIs are essential to driving decisions based on historical and real-time data. The fast pace of modern business necessitates the facility to be accessible to data at any point in time. It is essential for ABC Inc. to reduce IT dependency through applications run on the cloud, lowering the total cost of ownership

155

© Vinayak Gole, Shreekant Shiralkar 2020
V. Gole and S. Shiralkar, *Empower Decision Makers with SAP Analytics Cloud*,
https://doi.org/10.1007/978-1-4842-6097-5_6

(TCO) of the Analytics platform. In view of growing data security-related challenges, ABC Inc. also needs to ensure that data is secured from any potential business intelligence security issues by keeping that data safe and secure at all times.

Lacking "Mobility"

With mobile devices outnumbering the human population, gone are the days when producing work, planning, sharing information, or collaborating on projects was reduced to meeting rooms or desktop PCs alone. In a fast-paced, cutthroat digital economy, business leaders must be able to obtain access to data-driven reports and insights round-the-clock ubiquitously. A hyper-connected shift into the mobile age means that there has never been more demand for mobile-based BI solutions. At ABC Inc., most of the tools and applications are not mobile compatible and necessitate the use of PC's for dashboards and seamless access to office and outside. Meanwhile, due to the outdated technology on which these systems are built, it causes a lack of enablement on multi-device access and responsiveness. Also being tool based, they require access through specific tools as they are being accessible only over certain devices like desktops and laptops.

While developing mobile-optimized BI solutions can prove a developmental challenge, by working with the right interactive business intelligence platform, it is possible to log in and pull valuable insights from mobile devices from anywhere in the world without losing any key features or functionality. ABC Inc. is therefore keen to have an application that will be able to access trusted data in real time, even from sales reps' mobile devices. There's a need for increased ability and agility to help ABC Inc.'s customers by having a platform that improves decision-making, guiding product development, and helping customers to fuel customers' demands and enabling business growth.

ABC Inc. has envisioned their need to move to a modern Analytics landscape that should be the following:

1. Be available at all times,

2. Available across multiple devices,

3. Be responsive across the devices.

Alignment to Specific SAP Analytics Cloud Capability

As we have seen in the above sections, ABC Inc. has been losing out to competition due to the lack of a modern analytics platform. The current landscape at ABC Inc. is available only through a specific tool and can be accessed only within the organization's network. Hence crucial decisions take time, causing losing out on crucial business to the competition.

Some of the benefits of having the "Anytime Analytics" capabilities of SAC are the following:

1. **Accelerated decision-making:**

One of the major challenges that ABC Inc. faces in the current landscape is that of a longer wait time for executives to have access to data to make informed decisions. For example, a sales executive would have to rely on pre-prepared dashboards before a customer visit. Real-time access to data is not possible.

With anytime analytics, SAC enables executives to make real-time decisions. Decisions can be based on data and would not have to rely on pre-built datasets. Business leaders would be able to not just access data and reports from their mobile devices but also explore them in real time.

The mobile app for SAC also enables a strong cache management feature that ensures the data delivery is fast and readily available for rapid analysis.

2. **Actionable Intelligence:**

In the current scenario, Business Intelligence enables ABC Inc. to deliver reports to end users. These reports are standard in layout and primarily scheduled over the network or email. There is a limited amount of ad hoc reporting available as well. However, these features are not enough in the competitive world where actions need to be backed by not just data but also intelligence. Data and analytics are no longer only for presentations but influence everyday business decisions and prove to be a crucial component for future actions.

We learned in Chapter 5 how smart features and augmented analytics of SAC enable customers to have strong Actionable Insights. However, with mobility and cloud availability, these smart features can be used by the business community to rapidly interact with data from the benefit of a mobile device. The responsive screen enables access to this very intelligence to see the data in the right context and deliver the best possible results.

The mobile app also supports digital boardroom features. Executives can attend a board meeting remotely or even while traveling and have access to the same data being presented available on their mobile devices. Data-based intelligence is available at the swipe of a finger or from the comfort of a wider screen which makes the same analytics available for the best fit decisions.

3. **Rapid deployment:**

With SAC's Analytics Content Network, SAP provides multiple ready-to-use business content for rapid consumption. ABC Inc. can deploy these connections, dashboards, and stories with minor tweaks to enable end users to get immediate access to complex dashboards. This will improve the end-user experience and reduce the learning curve for end users who want to build dashboards for self-service BI.

One of the key tenets of SAP's strategy of developing SAC therefore has been to bring a robust analytics platform that should enable, not only advanced analytics to its end users but also be available at all times on any device. SAC is built as a cloud-native solution, that is, built from the ground up on the SAP Cloud platform. And since SAC is a cloud application, it is maintained by SAP, thus significantly reducing the maintenance effort for ABC Inc. Also, the cloud availability ensures availability at all times with no downtimes and round-the-clock access to data for analysis.

Let us now understand the capabilities of SAC that enable Anytime Analytics for Enterprises.

SAC Technical Capabilities

To understand SAC's capability of Anytime Analytics, let us first understand SAC Technical Capabilities.

The entire SAC architecture and the core technical capabilities, along with their positioning, in the landscape are shown in Figure 6-1.

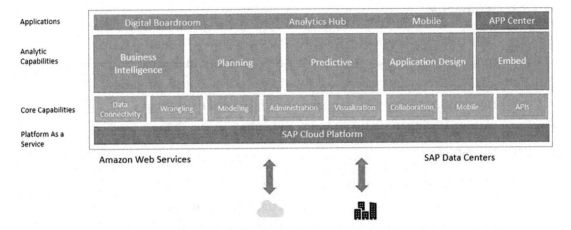

Figure 6-1. *SAC Architecture (Source: SAP)*

1. **Platform as a service (PAAS):**

SAC is built over the SAP Cloud Platform, which includes the in-memory HANA Database management system. The SAP Cloud platform enables robust applications to be built with multiple ready-to-use services. The robust HANA database ensures that data loading and wrangling is easily accomplished with rapid outcomes. Large volumes of data also can be easily wrangled over HANA. Since it is part of the SAP Cloud platform, SAC can also be used for embedded analytics over the S/4HANA Cloud platform.

2. **Core Capabilities:**

The data management capabilities of SAC are based on the HANA database platform, which enables fast in-memory processing and wrangling. The visualization capabilities are HTML5 based and hence can be rendered very effectively across multiple devices of varying screen sizes including mobiles. APIs are available for custom development of dashboards as well as connectivity to multiple systems.

The core capabilities include the following:

I. **Data connectivity:**

SAC enables live and import connectivity to multiple data sources – both SAP and non-SAP – which form the primary core capability. Data connectivity is an essential component of any analytics landscape, which should enable connectivity to a wide variety of sources. SAC provides native connectivity to multiple data sources – cloud and on premise.

II. **Wrangling:**

Data transformation or wrangling can be achieved through models and to some extent over stories. Models can be linked, and stories can be built across multiple models. Data wrangling capabilities enable SAC to easily merge data across sources without having to set up complex ETL jobs.

III. **Modeling:**

Modeling enables not just transformation of data but also ensures the quality of data is met, as well as enabling a semantic layer for end users and developers to work on creation of stories and dashboards.

Modeling capabilities also enable end users to develop their own understanding of data and contextualize it with business knowledge.

IV. **Administration:**

Any landscape without capable administrative capabilities does not come out to be very effective in managing. Even though SAC is fully maintained by SAP, administrators ensure the SAC landscape is effectively monitored and administered to enable a seamless experience for end users.

V. **Visualizations:**

Visualizations play a very important role in discovering trends and underlying data implications during analysis. Data visualization is an important aspect of SAC that provides multiple charts and graphs to enable end users to easily analyze the data. SAC also enables support for R visualizations for in-depth data discovery and analysis.

VI. **Collaboration:**

End users do not have to rely on third-party applications and tools to collaborate but can utilize the robust collaboration within SAC to share content and even chat over activities.

VII. **Mobile:**

Mobile internet is no longer a luxury and offers convenient and easy access to the World Wide Web. Mobile apps optimally use resources available in the handheld devices to ensure the best quality is delivered with the available hardware. SAC's mobile apps for iOS and Android operating systems ensure that data is available to all types of users at all times.

VIII. **APIs:**

APIs or Application Programmable Interfaces enable developers to build upon the existing capabilities of SAC into custom applications. SAC enables a strong API environment for developers to extend standard capabilities.

3. **Analytic Capabilities:**

Analytic Capabilities are the capabilities of SAC, which enable all analytics on a single platform. These capabilities form the frontline layer that enable data discovery as well as planning and predictive. These capabilities are built on services within the SAP Cloud platform that have been brought together to deliver a unified experience to the end user.

These capabilities include the following:

I. **BI or Business Intelligence:**

Business Intelligence forms the core of data analysis and enables building of models, stories, visualizations, and other components that are used for analytics and trend analysis.

II. **Planning:**

Enterprise planning enables organizations to budget and plan for the future. SAC allows all planning capabilities including allocations, spreading, and creating versions. Versions can be stored as private until the time they are not approved and then be shared with collaborators as public versions once they are approved.

III. **Predictive:**

Predictive capabilities enable end users to build predictive scenarios that can be used to forecast future trends. SAC currently provides three easy-to-create and deploy-predictive scenarios: regression, classification, and time series forecasting.

IV. **Application Design:**

Application design enables developers to build custom applications for specific requirements. SAC provides an application designer with support for scripting to enable developers to build and share applications.

V. **Embedding:**

SAC stories and applications can be easily embedded into other applications and websites either through APIs or through direct links. Apps created using APIs can be downloaded or purchased via the SAP App center as shown in Figure 6-1.

SAC enables tight integration between all of the above features and enables stories or applications to be built using a combination of one or more of these features.

4. **Applications:**

This is an additional layer built over the analytics layer to deliver further value to the end user. In Chapter 4 section, "Digital Boardroom," we learned about the SAC Digital Boardroom. The SAP Analytics Hub forms another web-based application that enables multiple analytical tools to be unified into a single platform so that they can all be explored from SAC.

The mobile app that we will be exploring shortly delivers the mobile experience, and the app center allows developers to build applications to be built or embedded into other applications.

We have now learned about the basics of SAC architecture and how SAP has designed architecture to enable cloud-native and responsive analytics. Based on this knowledge, let us now learn about the features that enable SAC to deliver "Anytime Analytics."

SAC "Anytime Analytics" Features

SAC enables end users and executives alike to be able to access data from any device at any point in time. Let us learn about the distinct features of SAC that enable "Anytime Analytics":

1. **Native to Cloud:**

SAC is a cloud-native application built over the high-performance SAP HANA Cloud platform. As we have already learned in Chapters 3 and 4, SAC enables a single version of truth and provides a single platform for all analytics. The entire application can be accessed through a web browser without the need for installing a local application.

Since there is no local installation, SAC can be accessed from a device with any OS. The browser also enables debugging the application intrinsically without the need for an additional application. The browser thus plays a very important role while working with SAC. Google Chrome and Microsoft Edge are the recommended browsers for working with SAC since these two browsers support all capabilities and features. Internet Explorer 11, though supported, provides view-only capabilities for SAC, which enable end users to consume analytics but does not provide the capability to create anything new. Internet Explorer 11 does not provide capabilities to create models or stories, use the digital boardroom, data explorer, or administer users.

Below are the reasons for having Chrome or Edge as the recommended browsers:

1. Better performance with Java-intensive applications.

2. Follow web standards, which ensure compliance for future versions.

Also, the entire landscape, including regular updates, is maintained by SAP, and enterprises do not have to spend time and effort in upgrading.

2. **Mobile apps:**

SAC provides a mobile app for both of the most popular mobile platforms, viz, iOS as well as Android. The iOS app has been available and has matured over the years whereas the Android app was launched in Q1 2020. The mobile functionality is especially handy for roles that need to have real-time access to data and analytics. The mobile apps are also customized to run efficiently with the mobile OS and deliver the best visualizations and results within the capacity of the mobile device.

Business executives can look at dashboards during their travel and also explore data for deeper insights. Sales executives can have in-depth knowledge of the prospect they are visiting and can bring in personalized negotiations as needed.

3. **Content Network:**

SAP provides pre-built out-of-the-box content for most Lines of Business (LOB)s and industries, which can be rapidly deployed through the Content Network. This enables end users to rapidly consume pre-built dashboards and stories from any location without the need to set up a separate development. SAP provides guides and content packages that can be downloaded and used with minor tweaks for establishing connectivity. Currently the packages supported are listed in Figure 6-2, with SAP adding further content regularly. The updated list can be accessed from `https://www.`

`sapanalytics.cloud/learning/business-content/`

LOB Packages	Industry Packages
Advanced Compliance Reporting	Agricultural Origination, Trading & Risk Management Analysis
Business ByDesign (Finance and Procurement)	Banking
Environment, Health and Safety	Chemicals
Field Service Management Performance Dashboard (New)	Consumer Products
Finance	Customer Profitability Analysis
Finance Account Receivable – Invoice Payment Forecasting	Engineering, Construction & Operations
Finance Contract Accounts (FI-CA)	Health Care
Finance – Live based on Semantic Tags	High Tech
Financial Consolidation S/4HC (BPE)	Industry Innovation Kit – Leonardo Zero Waste
Financial Planning & Analysis S/4HC (BPE)	Insurance
Financial Products Subledger IFRS17 for S/4HANA	Mill Products
FI Operational Expense Planning	Mining
Goods and Services Tax GST Analytics	Oil & Gas
Human Resources (SuccessFactors)	Professional Services S/4HC (BPE)
Human Resources (SuccessFactors) – Simplified Chinese Localization	Public Sector
Human Resources (SuccessFactors) (BPE)	Public Services: Higher Education and Research
Human Resources Salary Planning	Real Estate
Integrated Financial Planning for SAP S/4HANA	Retail (Model Company Fashion for Vertical Business)
Liquidity Planning S/4HC (BPE)	Retail (Model Company Core Retail)
Marketing	Retail - Omnichannel Article Availability and Sourcing
Manufacturing S/4HC (BPE)	Rural Sourcing Management
Procurement	Utilities
Procurement S/4HC (BPE)	
Product Cost Planning	
Project and Portfolio Management	
Project Budgeting & Planning S/4HC (BPE)	
Project Staff Planning	
Receivables Management for SAP S/4HANA Cloud (BPE)	

Figure 6-2. *SAC Standard Content*

We have learned how SAC's distinct features enable "Anytime Analytics" over the web browser as well as native mobile apps over iOS and Android.

Let us now learn the step-by-step process to understand how SAC enables anytime analytics capabilities for end users.

Enabling Anytime Analytics with SAC

As we have discussed earlier in this chapter, Anytime Analytics is one of the primary issues facing ABC Inc. The enterprise plans to modernize their Analytics Landscape to enable end users to be able to have anytime access to data from anywhere. We have also learned how SAC enables Anytime Analytics, primarily due to its cloud-native architecture, custom application development, and mobile apps.

In this section, we will learn how each of these components can be used effectively to deliver analytics directly to end users.

Accessing SAC over the Internet

As we have already learned in the section "**SAC Anytime Analytics Features**," SAC is a cloud-native application, and there is no separate application that needs to be installed but can be accessed over any web browser.

Let us now learn the step-by-step process to access SAC over the web.

> **Step 1: Request access:**
>
> SAC access can be requested to the SAC Administrator, who will provision access through the security module. Once access is granted, a welcome email is sent to the user with an URL to access SAC. Figure 6-3 shows the welcome email that is received by the requester once access is provisioned.

Figure 6-3. *SAC welcome email*

Use the URL provided in the welcome email to activate the account. Once the account is activated, it can be accessed over any device via a web browser. The access URL is the SAC cloud tenant deployed and assigned to a particular organization for building analytical components and data exploration. This tenant, though maintained by SAP, can be managed by SAC Administrators to enable user access.

Step 2: Logging in:

On clicking the URL provisioned in the welcome email, the login screen comes up. The login screen is shown in Figure 6-4. Enter the username and password to gain access to the SAC system. This very URL can be accessed from the web browser of any device, allowing seamless data discovery.

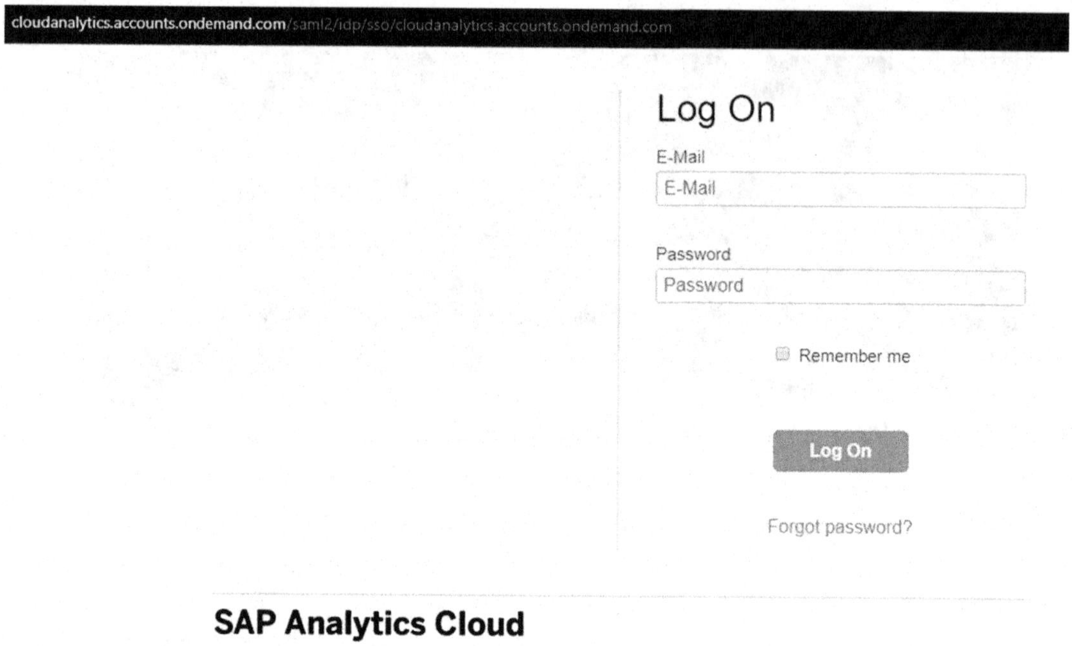

cloudanalytics.accounts.ondemand.com/saml2/idp/sso/cloudanalytics.accounts.ondemand.com

Log On

E-Mail

E-Mail

Password

Password

☐ Remember me

Log On

Forgot password?

SAP Analytics Cloud

Figure 6-4. *SAC login screen*

The application adjusts automatically to the screen size and delivers the very same experience as on a larger screen device.

In this section, we learned about how to access SAC across any device over the internet to enable end users to access SAC using any web browser.

Let us now learn the step-by-step process to install the mobile app and access it to explore data and create analytics.

Connecting to SAC Using a Mobile App

SAC is available over both popular mobile OS devices: iOS as well as Android. The mobile app can be downloaded from the Apple Store or from the Google Play Store. For demonstration, let us consider the Android app as an example. Before using the mobile app, install the app as shown in Figure 6-5.

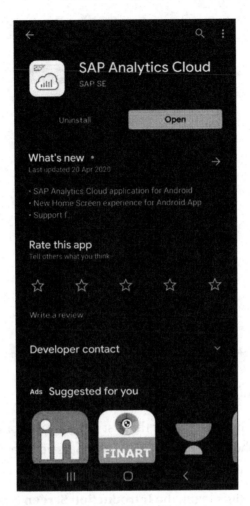

Figure 6-5. *SAC Android app on Google Play Store*

Go to the Google Play Store and search for the SAP Analytics Cloud app. Install the app on the mobile device. Once the application is installed, click on the shortcut, which will launch the app. Enter the URL of the tenant where to connect to, and this will launch the app as shown.

Once the app is installed and set up, follow the steps as shown in Figure 6-6 to explore analytical content directly on the mobile app.

Figure 6-6. *SAC Android mobile app*

Step 1: On opening the app for the first time, a password needs to be created as shown in Figure 6-6. Set the password, post which the password can be used to log in.

Step 2: For first time login, the **Introduction Screen** comes up. Post first-time login, step 3 comes up post login without the need for the Introduction to the app. Click "**Continue**" to move to the next step.

Step 3: The SAC **File Explorer** comes up. This is in conjunction with the web-based SAC User Interface (UI). The UI is intuitive and can be explored just like the web application. SAC Content can be searched or explored through the **File Explorer**.

Step 4: Drop down to the **Public Folder** and subsequently to the folder where the files to be explored are stored. Let us open the **Story "Demo for Smart Discovery."** This will help us compare the output of the story on the mobile device as well as the desktop.

Step 5: The **Story** comes up as shown in Figure 6-6. The responsivity of the SAC canvas has enabled the story to be fit into the screen size available. It can be further scrolled down to explore the graphs, charts, and other information.

The subsequent pages can be navigated by swiping across the screen in either direction as shown in Figure 6-6.

We have learned how SAC can be accessed over a mobile app and provides availability to the content over a mobile device. This enables another path of the "Anytime Analytics" of SAC. Due to the responsiveness of the screen, SAC content can be directly accessed via the browser on a mobile device as well. The responsive pages in the Story will be automatically resized, and data visualization will fit into the available screen size. Currently canvas and grid pages are not supported for mobile apps. To see which stories are mobile enabled, refer to the File browser, which shows the mobile symbol next to the story.

This is shown in Figure 6-7.

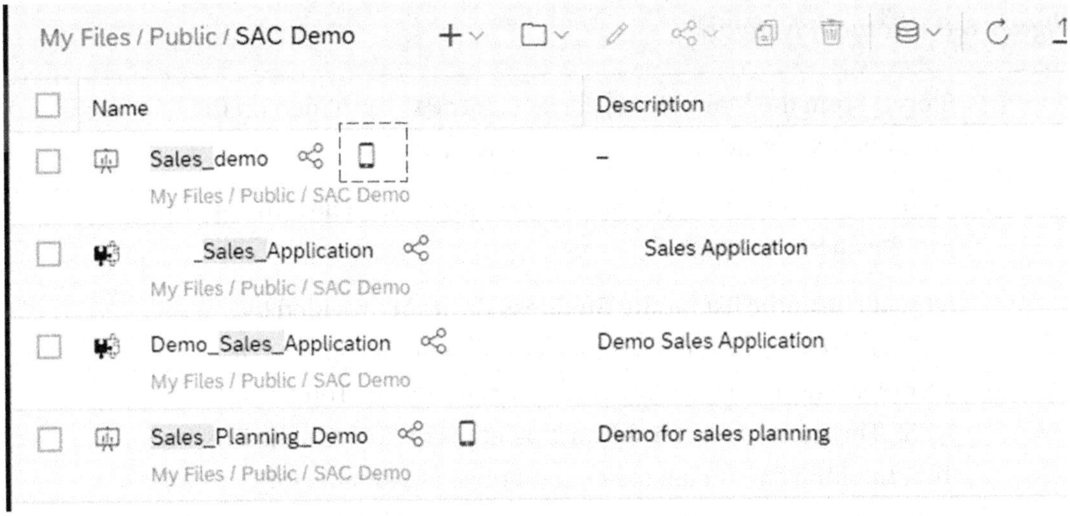

Figure 6-7. *Mobile-enabled Stories*

Deploying Content Network Analytics Content

Content Network enables access to **Business Content** provided by SAP to rapidly deploy SAC content. Business content provides pre-built content in terms of models, stories, and boardroom content for quick deployment and rapidly built analytics in SAC.

Let us now learn the step-by-step process to access the Analytics Content Network. The entire process is shown in Figure 6-8.

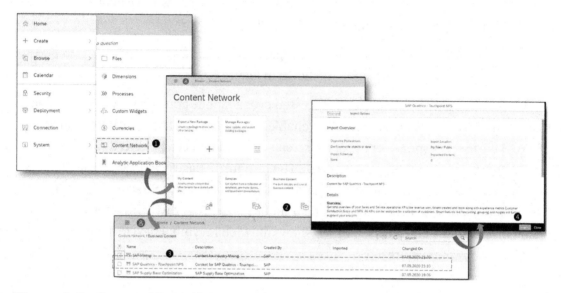

Figure 6-8. *Content Network*

Step 1: From the Main Menu, click on "**Browse**" and then on the **"Content Network"** as shown in Figure 6-8.

Step 2: The **Content Network** window comes up. Select the option for "**Business Content.**"

Step 3: The entire list for the **Business Content** provided by SAP comes up. Select the content as shown in the figure. For demonstration, we have selected "SAP Qualtrics – Touchpoint NPS." The content screen shows the details of "SAP Qualtrics – Touchpoint NPS" and all the content objects available for download.

Step 4: Click on "**Import**" to download the content to the SAC landscape. The downloaded content is available in the folder SAP_Content, from where it can be modified if needed and moved to the appropriate folders.

Let us now learn about the components of Business Content. The Business Content consists of two components:

1. Overview

2. Import Options

1. **Overview:**

   The "Overview" section of the content Import also provides links to detailed instructions on how to install the content and changes to be done to the connections as well as changes to be done in the source systems.

   This is shown in Figure 6-9.

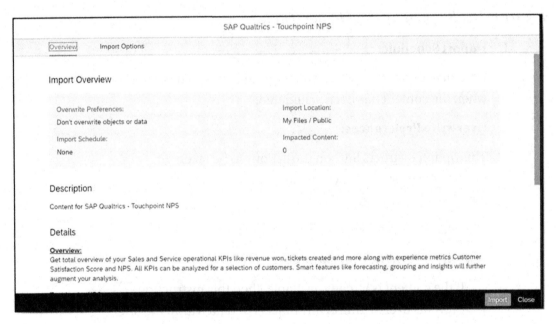

Figure 6-9. *Import Business Content Overview*

2. **Import Options:**

   While importing the Business Content from the Content Network, SAC offers options for importing content. The import options are shown in Figure 6-10.

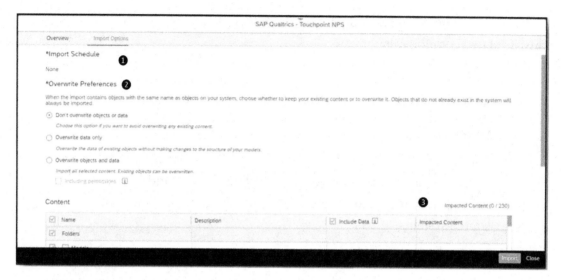

Figure 6-10. *Business Content Import Options*

1. **Import Schedule:**

   Content can be scheduled to be imported. This option shows when the content has been scheduled.

2. **Overwrite Preferences:**

   This option instructs SAC on the actions to be taken for overwriting existing objects, including models, stories, dimensions, and connections.

   The available options are the following:

   a. **Don't Overwrite objects or data:**

   If the content is already available, then this instruction prevents overwriting of any objects or data within the SAC landscape. If the standard business content is already in use, this option enables accidental overwriting of the content.

   b. **Overwrite data only:**

   This instruction enables SAC to only overwrite the data from the business content and the objects are not overwritten. This instruction thus updates only the newer data and the objects that are already in use are not overwritten, preventing any change to the objects being used by end users.

c. **Overwrite objects and data**:

This instruction enables SAC to overwrite all objects present in the landscape as well as newer data from the business content. This option is useful while doing a complete refresh and enabling newer content to be available for end users.

3. **Content:**

This option shows which of the existing objects are affected by importing the new content. It also provides an option to select specific objects to be overwritten. This list is extremely useful to determine if any frequently used object is being overwritten or affected and to prevent further problems.

In this section, we have learned how to set up SAC to be accessible at any time using the web application, and we have also learned the step-by-step process to explore data over the mobile app.

We have also learned about standard content available and how it can be used to rapidly enable anytime analytics for end users.

Summary

In this chapter, we have learned how data and information are available in real time, at reduced IT dependency through applications run in the cloud, while data is secured from any potential analytics platform security issues. We have learned how SAC can be used to deliver a seamless data experience over any device with minimal effort. We have also seen how the mobile apps enable access to SAC over any device. We have also seen the step-by-step process to set up the Android mobile app and explore content over the SAC landscape. We have further learned about content network and how it enables rapid deployment of content. We finally understand how these features enable "Anytime Analytics" enabling enterprises to make crucial decisions in real time.

In the next chapter, we shall learn how SAC's predictive analytics featuring Smart Predict can enable rapid deployment of predictive scenarios for advanced forecasting and predictive capabilities.

CHAPTER 7

Capitalize on Predictive Analytics Capability through SAC

With a view to being competitive, ABC Inc. is keen to identify new revenue streams while consolidating its existing revenue base by identifying upselling and cross-selling opportunities of their existing portfolio of products and services. Predictive Analytics has been identified to support this endeavor; however, their best efforts and attempts to apply predictive analytics have met with very minimal success.

Citizen data scientists and opportunities to enable business users with the power of data science have proven detrimental to the organization, with competition gaining the upper hand. Market share has been slipping due to lack of direction in forecasting and planning initiatives. Moreover, business users with deep business knowledge are restricted to using standard reports that provide limited capability for data handling and mining.

As cited earlier in Chapter 2, rapid business growth of ABC Inc. as well as their engaging with acquisitions aggressively have resulted in huge diversity in information and systemic inability to integrate data sources, creating difficulty in capitalizing on the vast data. But the basic challenge posed is that this data is spread across a variety of disparate systems and software and stored in various ERP systems, CRMs, databases, and Excel spreadsheets. With data spread across multiple systems, getting the information is an arduous task, and deriving insights that could help them decide for the new revenue sources or upselling and cross-selling opportunities remain a distant dream. Questionable data quality, compromising reporting accuracy, forces a huge effort in reconciliation, and traceability is a major challenge. Decision support therefore

© Vinayak Gole, Shreekant Shiralkar 2020
V. Gole and S. Shiralkar, *Empower Decision Makers with SAP Analytics Cloud*,
https://doi.org/10.1007/978-1-4842-6097-5_7

consumes an enormous amount of attention and decision-makers often lack confidence in the results. Attempts to derive gain from Predictive Analytics therefore haven't been encouraging at ABC Inc.

It has been recognized that ABC Inc. is facing multiple challenges due to an outdated system. We appreciated some of these challenges while learning about augmented analytics in the previous chapters. One of the major challenges that ABC Inc. faces due to the unavailability of a modern analytical system is the lack of predictive analytics. Data Science initiatives are restricted to specialists, and there is a complete lack of explainable AI within the current tool set the organization houses.

While improved data quality allowing users to trust the data for decision-making is paramount, management at ABC Inc. aspires to have a solution with the capability to consolidate data from diverse operational units into a single business intelligence reporting platform at the pace of the business, instead of business being chained by their systemic limitations. They also need to provide greater insights with capabilities like Predictive Analytics to end user teams across the world.

ABC Inc.'s requirement is to have a modern analytics platform that will provide users with a self-service data science mechanism. This could be in the form of ready-to-use algorithms with a simple interface for providing a guiding mechanism for forecasting and decision-making. This should further be part of the **"All-in-One"** analytics platform and should integrate with other analytics functions like Business Intelligence and planning. This platform will enable users to create all data mining activities within a single platform and save time in terms of working with multiple technologies, interfaces, and tools.

Citizen data scientists and end users with deep domain knowledge can add contextual information to the predictions made by the algorithms and provide business decisions with the competitive edge. For example, tweaking the supply chain by right-time inventory management can deliver exceptional cost savings with minimal wastage. Similarly, the Planning functions can be augmented by the predictive feature to aid organizational planners to organize plans for the best utilization of resources and enable just-right planning. The predictive analytics features will also ensure that the data is always aligned to the **"Single Version of Truth"** principle. Thus, enabling a predictive analytics platform with minimal investments and a low learning curve could improve the competitive edge of ABC Inc. considerably.

In the next section, let us learn how SAC provides an integrated solution for Data Science and Predictive Analytics Capability.

Alignment to Specific SAP Analytics Cloud Capability

In the previous section, we learned about the challenges and difficulties that ABC Inc. has been having while attempting to integrate Predictive Analytics Capability into the analytics platform of ABC Inc. We have also appreciated the importance of having a predictive analytics capability for building a competitive edge. Augmented Analytics powered by predictive capability allows end users to adopt analytics in everyday business decisions and tasks increasing efficiency across the organization. Powered by data and explainable trends, each decision can be backed by a robust and collaborative system with specific rules, leaving little room for errors.

SAP has integrated predictive capability within the SAC landscape by allowing end users to create predictive scenarios and integrating those scenarios into analytical and planning stories. SAC delivers a custom-built predictive analytics capability solution with Smart Predict. Integrating the features of predictive analytics, the platform delivers a solution incorporating a collection of algorithms and powered by machine learning selection for best fit.

Smart Predict allows end users with a black box predictive analytics platform wherein the end user has no need to delve deep into the complex predictive model creation process. With ready-to-use scenarios and the capability to evaluate the best fit model based on data and inputs provided by the end user, Smart Predict delivers machine learning-powered predictive analytics capability with the ability to easily implement the same.

SAC's Smart Predict enables end users to incorporate predictive analytics into everyday data analysis to power improved business decisions and outcomes. Smart Predict along with other augmented analytics capabilities of SAC known as "Smart Assist" encourage business analysts within the organization to mine data and contextualize with business domain knowledge to drive improved and actionable insights. We have already learned about Smart Assist and its components in Chapter 5.

Here are some of the benefits offered by Smart Predict:

1.   Informed Decisions:

   Key decisions across organizations are typically based on end-user experience. With the emergence of Business Intelligence, key decisions started being driven by data-backed Key Performance Indicators. Predictive analytics capability has pushed the needle further by enabling end users to have an informed view about the future rather than just about the present. Statistical analysis has proved to be the edge in a competitive market.

   SAC's Smart Predict technologies enable not only statistical analysis but provide end users with an easy-to-use explainable technology tool, with a low learning curve. Enabling end users with Regression for predicting numerical values, Classification for binary decisions, and Time Series for forecast over time, Informed decisions can help organizations plan for the future.

2.   Reduced Risk:

   Smart Predict enables automated analysis of historical data with simple explainable technologies to enable end users to make Informed Decisions. Since the decisions are now driven by data and not on past experience, the probability of a disruption is very minimal.

   Also, Smart Predict enables analysis of trends, cycles, and fluctuations. This can help key decisions be based on a higher probability of an occurrence rather than mere hunches. Investments can thus be targeted for specific use cases, thus considerably reducing the risk involved.

3.   Improved Business outcomes:

   With decisions backed by powerful analytics and reduced risk, business outcomes can be improved considerably. With business plans targeted specifically for signal values based on forecasts, investments can be made specific to business lines and products ensuring business outcomes. For example, the

Time Series example we have seen in the previous sections will enable us to get a forecast of the sales view. With a specific target, the organization would be able to plan accordingly and create a business plan to match. Classification can be used to target specific users to create a personalized marketing plan. With specific predictive scenarios, organizations can drive better outcomes for business plans and improved investment results.

Before we learn further the capabilities of Smart Predict, let us now learn about the basic components of implementing predictive modeling algorithms in SAC.

Overview

Smart Predict provides the end user with ready-to-use Predictive Scenarios that can be rapidly deployed on historical data to deliver recommendations for the next best course of action. Smart Predict augments the Business Intelligence capabilities of SAC and is an important component of the Augmented Analytics features of SAC we explored in Chapter 5. Smart Predict also provides predictive capabilities built into a single platform, and the recommendations provided can be rapidly built into stories for enabling end users to get a view of possible options for future conditions, thus enabling informed decisions.

Note   Smart Predict is available as a separate license.

Before we learn about creating models with the Smart Predict feature, let us understand the primary components.

The primary components of the Smart Predict feature are the following:

1. Dataset

2. Predictive Scenario

3. Predictive Model

Let us now learn about each of these components in detail.

Dataset

A dataset can be considered similar to a table with data arranged in a columnar structure. The dataset is more aligned to a particular predictive model, and each of the columns represents a variable. The primary objective of a dataset is to help the model with the appropriate data at each stage of its creation and execution.

Datasets can be categorized into three types:

- Training Dataset

- Application Dataset

- Output Dataset

Let us learn about each of these datasets in detail

Training Dataset

Data in the Training type of a dataset is used to train the model and build the predictive scenario. Each scenario is built for a particular type of business problem, and the data has to be sanitized accordingly to train the model optimally. This dataset is typically historical and allows the model to understand the relationships between data components and put out variables that define the reliability of the model in accurately predicting the outcome for a particular dataset. A model can be trained with multiple sets of data to determine the best form of data to be provided to the model for ingestion.

The size of the dataset used for training the model should be of sufficient size to enable the model to predict the outcome accurately. If the sample size is smaller than thousands, the model might be influenced by variation in the dataset or noise. Especially for a Classification Predictive Scenario, the data should also have a wider variety of records to factor in for all possible cases in the dataset. This would enable the model to be refined further and deliver a strong Predictive Model.

Application Dataset

The data in the Application type of dataset is the input to the trained predictive model. Once the trained model has built the relationships and the best possible coefficients, it is able to predict the best outcomes on the application dataset. The application dataset has to be in tune to the training dataset that delivered the best results from the training model. The Application dataset is typically current and uses future data on which the trained model will be used to predict the outcome.

The Application dataset thus has to follow the same quality standards as the training dataset.

Output Dataset

The output dataset is the output of the predictive model that has processed the data from the Application Dataset and delivered prediction. The delivered prediction output can be acceptable or be rejected based on the values of the variables. Once the output dataset has been confirmed to meet standards, it can be exported to create a model on which a story can be built. It can also be used as an input to other applications for forecasting and integrated reporting.

Predictive Scenario

A predictive scenario is a workspace set up to define a particular forecasting problem or use case. The predictive scenario in SAC allows building of a focused predictive model for a defined set of variables. These variables also offer comparative analysis to isolate the best set of results. SAC currently offers three of the most common predictive scenarios.

Time Series

The Time Series predictive scenario is used to forecast future values based on time as the base. The forecast will depend on the underlying model's interpretation of cycles, seasons, time, and other variables that would influence the future value.

Some of the common examples of time series forecasting are the following:

  i.   Group Profit forecasting over a year.

  ii. Revenue forecast across each month over the quarter.

  iii. Costs and expenses over the next two months.

Classification

A Classification predictive scenario is primarily used to determine the likelihood of an event occurring in the future based on information from past data. Influencing variables determine the probability of an event occurring with the same influences that have occurred in the past.

Some of the common examples of Classification are the following:

i.   Targeted promotions based on customer purchasing history.

ii. Employee attrition segmented by demographics or age.

iii. Fraud detection based on customer spending behavior.

Regression

A Regression predictive scenario is used to predict a numerical value for a variable based on other influencing factors. Influencing factors are described as input to the model that predicts the expected value for the target variable based on the historical values.

Some of the common examples of Regression are these:

i.   Price forecasting for a particular product based on demography.

ii. Target customers visiting a website based on past website analytics.

iii. Expected sales for the next month.

Predictive Model

A predictive scenario can be considered as a collection of predictive models. A predictive model is the result of data analysis brought forth by SAP automated machine learning. Once a predictive scenario is selected and inputs provided in terms of a target variable and training dataset, the machine learning algorithm mines the training data to find the best fit relationships and the variables that can deliver the most relevant results. This outcome of training is considered as the predictive model. If the results are not satisfactory, then the variables can be changed or a different training dataset provided until the Smart Predict Scenario delivers a model best fit for the particular business problem. Once the training of the model is completed, it can be applied to the target dataset to derive predictive responses to business queries or trends for future time periods.

Variables and Roles

Variables are the columns of the dataset made available for the predictive model. Variables have to be assigned specific roles to ensure that the output of the predictive model is as desired and the best fit for the business query put forth. Smart Predict predictive models have the following list of variables included.

Target/Signal Variable

This is the variable on which the output of the predictive model is based. This is also the answer provided to the business query put forth to the model.

Date Variable

This is the date on which the data is based. This is one of the most important components of a Time Series model.

Segmented Variable

This variable enables division of data into segments enabling the Target/Signal Variable to be customized on these segments.

Excluded Variable

This option decides which columns of data to be excluded from the predictive model.

Influencer Variable

This is the contextual information that is used to support the output provided by the predictive model.

In this section, we have learned about the primary components that make up a predictive scenario. An overview of these components will help us learn the Smart Predict application better and aid our learning in the next section, where we will learn about the step-by-step process to create a predictive scenario.

Step-by-Step Process to Build a Predictive Scenario

In the previous section, we learned about the different components of Smart Predict. Before a predictive scenario can be created, the underlying predictive model needs to be trained with the appropriate training dataset that holds historical data of two years. Subsequently we will learn how to configure settings for achieving optimal and best fit results from the predictive models. Let us first learn the step-by-step process to create a dataset. Post training, the model needs to be applied to the application dataset. The same process is applicable for creating both types of datasets. For the purpose of demonstration, we will create a training dataset and apply it to a model to predict the future trend. We will consider the Time Series dataset for the purpose of demonstration.

As we have seen in the previous section under **"Predictive Scenario,"** an essential variable for a Time Series predictive scenario is a date. We need to ensure that date is a component of the dataset we would be creating.

Step-by-Step Process to Create Dataset

As we have understood in the section **"Dataset,"** a dataset is a is a collection of data arranged in a columnar fashion. The dataset should be aligned to the type of predictive scenario selected. Since we are building a Time Series Predictive Scenario, the dataset should have a date or time dimension to ensure the predictive model delivers the best results as per the expectation. Initially the dataset would be built for training the model, and subsequently the actual dataset would be used for gaining predictive information from the selected predictive model.

Let us first learn how to create a dataset. The entire process is shown in Figure 7-1.

Figure 7-1. *Step-by-Step Process to create Dataset*

Step 1: From the main menu, click on **"Create"** and then on
"Dataset" as shown in Figure 7-1.

The **Create Dataset** window comes up, which presents two
options for creating a dataset:

a.   *File:*

With the **File** option, a file can be uploaded into SAC and can
be used as a source for training the model or for applying a
trained model to the dataset. The file can be uploaded from a
local computer or a shared drive.

b.   *Data source:*

A **data source** is a source of data that can be connected to by
a connection. Once an import connection is created, a set of
source data can be extracted and loaded into SAC as a dataset.
The data ingested can then be processed further.

Click on ⬜ to select data from a file as shown in Figure 7-1.

Step 2: The file selection screen comes up as shown in Figure 7-1. Select the file "**Sales_data_Sample_Timeseries_New**" from the local computer.

Click **Import** as shown in Figure 7-1.

Step 3: The **Save** window comes up. Let us give an appropriate name to the dataset as well as a description. In this case we have named the data set **Sales_data_sample_Timeseries_New** as per the name of the file uploaded. Click **"OK."**

Step 4: The dataset is created as shown in Figure 7-1.

In the previous subsection, we have learned to create a dataset. Let us now learn how to create a predictive scenario based on this dataset. For the purpose of demonstration, we will use the Time Series scenario. A similar process can be applied to regression and classification predictive scenarios for building models. Selection of the predictive scenario is primarily dependent on the business case. We have discussed the predictive scenarios and the business use cases in the previous section on Predictive Scenarios.

First, let us learn about the Predictive Scenarios Application screen and the components.

The Predictive Scenarios Application

We have seen in the previous section how a dataset can be created. The next step is to create a Predictive Scenario. The Predictive Scenarios Application enables end users to define a business problem for predictive technologies and deliver recommendations on the problem defined.

SAC Smart Predict provides three built-in predictive scenarios: Classification, Regression, and Time Series. The Predictive Scenarios Application is shown in Figure 7-2.

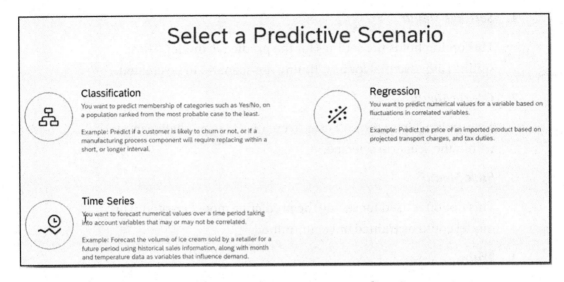

Figure 7-2. *SAC Smart Predict Predive Scenarios Application*

On selecting any of the predictive scenarios, the Predictive Model Creation window comes up. This is shown in Figure 7-3.

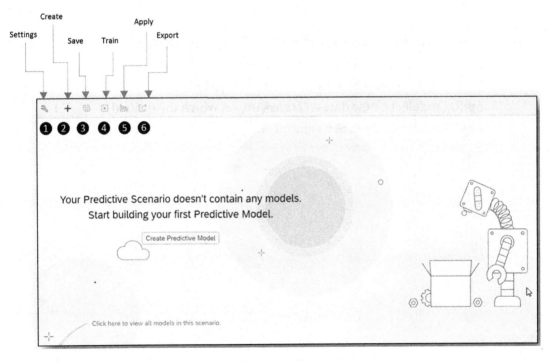

Figure 7-3. *Create Predictive Model*

1. **Settings Menu:**

   This option holds the settings for the predictive model. These settings are essential for fine-tuning the scenario to be created.

2. **Create Menu:**

   This option allows the end user to create a new predictive model within the scenario selected.

3. **Save Menu:**

   This option is used for saving the predictive model created. This model could be trained or yet untrained.

4. **Train:**

   Once the model is created, it can be trained with the right dataset. This option is used for training the predictive model. We will learn more about training the model when we understand the process of creating a predictive model.

5. **Apply:**

   Once the model is trained with historical data, variables fine-tuned, and the model finalized, it can be applied to future datasets to initiate the prediction. By clicking the **"Apply"** option, the option to select a new dataset comes up, on which the trained model can be applied. See Figure 7-4.

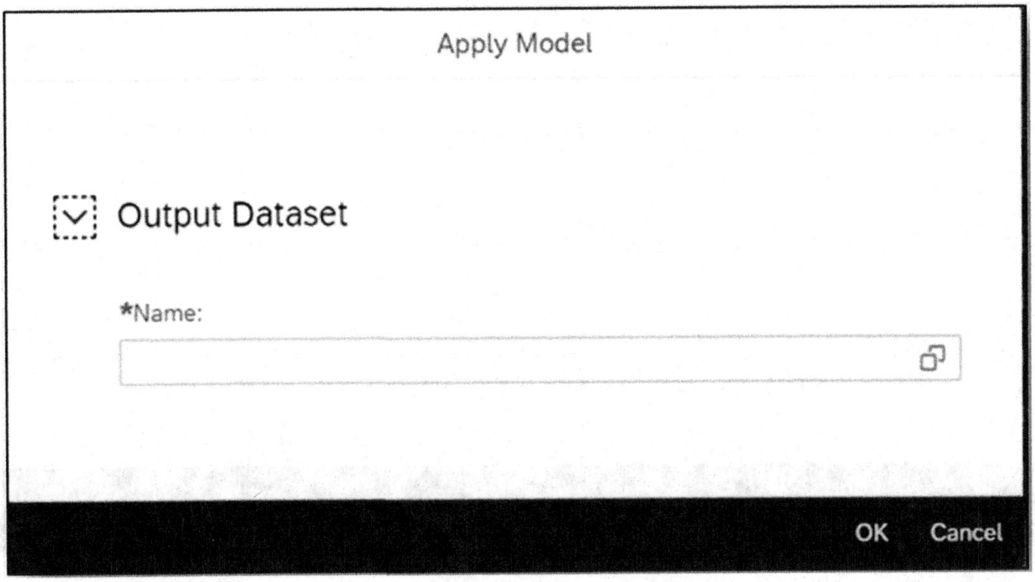

Apply Model

☑ Output Dataset

*Name:

OK   Cancel

Figure 7-4. *Select output dataset for Predictive Model*

6. ***Publish:***

SAC allows the Smart Predict model to be exported to other
applications such as SAP S/4HANA. With the Predictive Analytics
Integrator (PAi), the trained models can directly work on data
from other applications without the need for data to be uploaded
to the SAC landscape. In a highly integrated environment, SAC
can be used as the predictive analytics capability tool of choice,
and the created models can be dissipated to other applications.
Figure 7-5 shows the Publish menu.

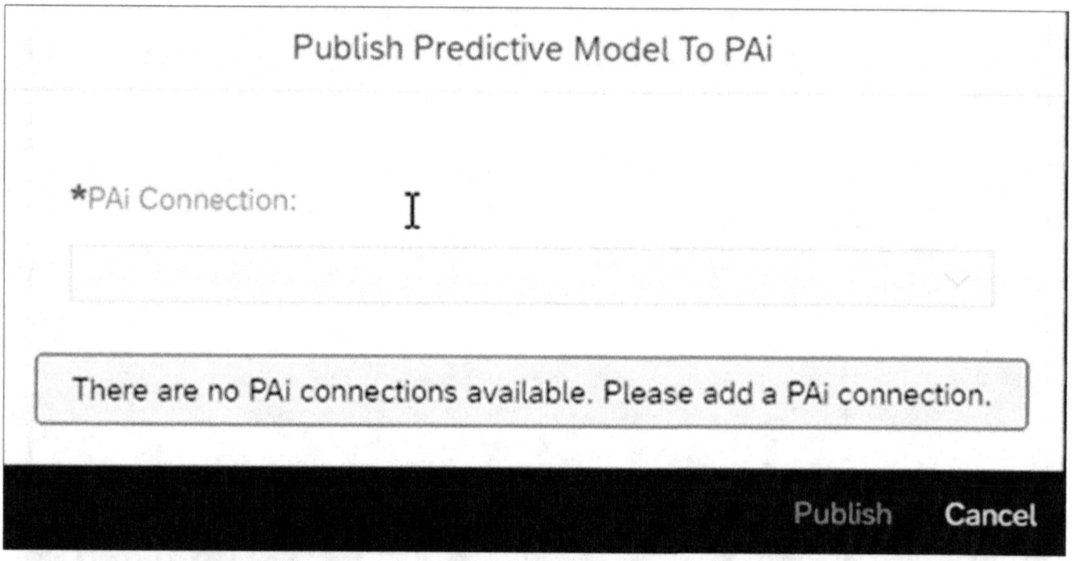

Figure 7-5. *Publish Predictive Model to PAi for external consumption*

We have learned about the various menus available on the Smart Predict model creation screen. While creating a model, there are multiple options that need to be set in terms of defining variables and segments.

Let us learn about the predictive model settings in detail.

Predictive Model Settings

The Smart Predict model that is created has to be fine-tuned by setting the appropriate variables and settings for influencers as well as training settings within the SAC environment. This will ensure that the output of the predictive model is optimal and a best fit for the business problem at hand. To ensure that the predictions or recommendations delivered are aligned to the business problem, the settings play a crucial role and have to be ensured they are set correctly.

Figure 7-6 shows all the Predictive Model settings. Let us learn each of these settings in detail.

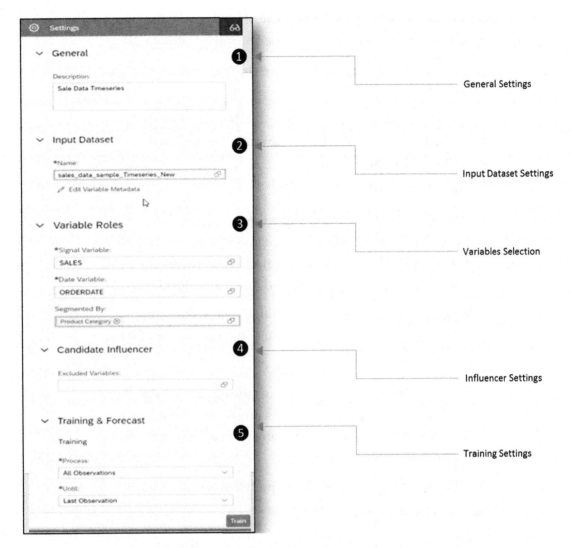

Figure 7-6. *Predictive Model Settings*

1. ***General Settings:***

   General Settings display the information about the model being created as well as the description of the model. This is shown in Figure 7-6.

2. ***Input Dataset settings:***

   The Input Dataset settings section allows for selecting the input dataset based on historical values. This dataset is used for training the predictive model. This is an essential component of training the model since selecting the right data set will ensure the model is trained to the best of its capabilities. The Input Dataset settings are shown in the figure.

3. ***Variables Selection:***

The Variables Selection section allows for defining the variables on which the predictive model will be created as well as the supporting contextual variables.

a. *Target/Signal Variable:*

The target variable or the signal is the measure on which the prediction is built. In the case of regression and classification, this variable is known as the Target Variable whereas in the case of a time series model, this is known as a Signal Variable.

In Figure 7-6, we have selected **"Sales"** as the signal that indicates the prediction will be built on future values of sales.

b. *Date Variable:*

The Date variable is the underlying time dimension that is the contextual variable in a time series predictive model. For a time series forecast, this is mandatory since the entire forecast is based on time.

In Figure 7-6, this is depicted by the **"Orderdata"** variable.

c. *Segmented by Variable:*

A segment is a category into which the measure can be distributed into. For example, sales can be distributed by products or product lines. This enables the forecast to be distributed accordingly, and a customized forecasting can be built based on each category.

In Figure 7-6, **"Product Category"** is selected to define the Segmented by Variable.

4. ***Influencer Settings:***

As the name suggests, Influencer Variables are used to bring into consideration other data points that add further influence into the predictive forecast.

Figure 7-6 shows which influencer variables can be excluded to narrow down the time series forecast for the selected target variable with respect to time. We have not selected any variable to be excluded in Figure 7-6.

5. ***Training Settings:***

These settings define which values to consider for the training of the model specifically if all time periods need to be considered.

We have now learned in detail the components and settings of the Predictive Scenarios application. Let us now learn the process of creating and fine-tuning the predictive model.

Stages of Predictive Model Creation

In SAC Smart Predict, the creation of a predictive model consists of primarily two stages:

1. Training Stage

2. Debriefing Stage

Training Stage

The first stage of creating a model is the Training Stage. In this stage, the model is supplied with historical data to help it understand data relationships and comprehend how the variable selected reacts to changes in data.

While creating a predictive scenario, the internal algorithm of SAC creates multiple models based on the input dataset. Each of these models is evaluated to finalize the best fit model and display the output. SAC Smart Predict splits the training dataset into two components, 75% of which is used for training the model and the remaining 25% is used for validating the model output of the trained model. One of the salient features of Smart Predict is the feature wherein multiple models are evaluated rapidly and the best fit model presented to enable the end user considerable savings in time spent.

Debriefing Stage

In this stage the performance of the selected variables is evaluated over several parameters to decide if the model is ready to be applied for active datasets or to be trained and fine-tuned further. The debriefing stage is essential to bringing in the human inspection aspect once the SAC Smart Predict algorithm selects the best fit model. Human debriefing enables the model to be further tuned and tested, enabling end users to ensure the model delivered is the best fit solution to the business problem defined.

Now that we understand how SAC Smart Predict evaluates and creates the best fit model, let us learn the step-by-step process to create and train a model.

Step-by-Step Process to Create and Train Predictive Model

ABC Inc., which has interests in multiple businesses, would like to analyze their motor vehicles sales. To plan accurately for future investments, having an estimate for sales across multiple categories would enable organizational departments to spend resources to align with the plan. This would also enable analysts to determine the effect of external factors on sales. This would further enable the leadership at ABC Inc. to make informed decisions on vehicle sales and the strategy for future endeavors.

With SAC's Smart Predict capabilities, Predictive Scenarios would enable ABC Inc. to forecast sales across vehicle segments. The analyzed data can then be integrated with stories and plans including value driver trees, which we discussed in Chapter 4 under the section "Value Driver Trees." The planning team can also align plans and allocate resources as per the forecasts.

To facilitate the above, let us learn to create a Predictive Model based on the Time Series Scenario. As we have learned in the previous section on "Predictive Scenarios," a Time Series scenario enables forecasting a variable on the basis of time. In our example, the variable is **"Sales,"** which will be based on **"Orderdate,"** which is the time dimension.

The entire process for creating the model based on a Time Series predictive scenario is shown in Figure 7-7.

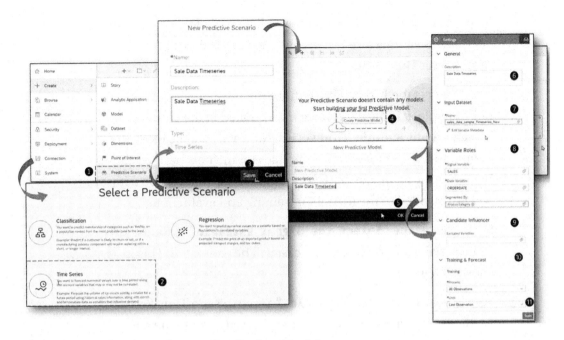

Figure 7-7. *Create and train Predictive Model*

Step 1: From the main menu, click on **"Create"** and select **"Predictive Scenario"** as shown in Figure 7-7.

Step 2: The **Select Predictive Scenarios** screen comes up. Three options are currently available for selection: Classification, Regression, and Time Series. We will be working with the Time Series Predictive Scenario; hence as shown in Figure 7-7, click on **"Time Series."**

Step 3: The **New Predictive Scenario** window comes up as shown in the figure. Name the new Time Series Predictive Scenario as shown in Figure 7-7.

We have named the Predictive Scenario as **Sales Data Timeseries** since we are forecasting Sales. Click **"Save."**

Step 4: The Predictive Scenario window comes up as shown in Figure 7-7.

Click on **"Create Predictive Model."** Note that we have created a Predictive Scenario in step 3 and now we are creating a Predictive Model. A Predictive Scenario can have multiple models.

Step 5: Enter a description for the predictive model and click on **"OK"** as shown in Figure 7-7. We have given the description as **"Sales Data Timeseries."**

Step 6: Select the **Input Dataset** from the settings window. In our example, it is the dataset we have created in the subsection on **sales_data_sample_Timeseries_New**. We shall be using the dataset to find the forecast for the sales for the next periods and hence will be using the Time Series model. The data in the dataset will be provided to the model for training and evaluation.

Step 7: Select the **"Select the Signal Variable"** as **Sales.** This is the variable on which the forecast will be generated. This is shown in Figure 7-7.

Step 8: Select the **"Date Variable."** Since this is a Time Series predictive scenario, the forecast is generated across time and hence the Date Variable is very important and hence mandatory. Select **ORDERDATE** from the dataset as the Date Variable for the forecasting of **Sales**. This is shown in Figure 7-7.

Step 9: If the forecast is to be calculated for categories within the data, then select the **"Segmented by"** Option. In our example, we have categorized by products and hence select **Product Category**. This will help create the predictive forecast based on **Product Category** and would enable objective evaluation of the forecast for each category.

Step 10: Since we are running for **All observations,** keep the default selection as shown in the figure.

Step 11: Click **"Train"** as shown in Figure 7-7. This will initiate the training of the model and bring up the results screen, which we will learn about in the next section.

We have now learned the step-by-step process of creating and training a model. Once the best fit trained model is presented by the SAC Smart Predict algorithm, the debriefing of the model has to be done by the end user to ensure that the results are best aligned to the business problem, which in our example is the forecasting of Sales for the next periods.

Debrief the Predictive Model

Once the model is created, it has to be debriefed to fine-tune and understand the validity of the forecast. A typical Debrief will consist of three sections:

1. Overview

2. Forecast

3. Signal Analysis

Let us now learn about each of the sections of the Debrief of the Predictive Model. The understanding of this section is essential to evaluate the validity of the model and then decide whether to retrain the model or to continue with the best fit model presented by SAC's Smart Predict algorithm.

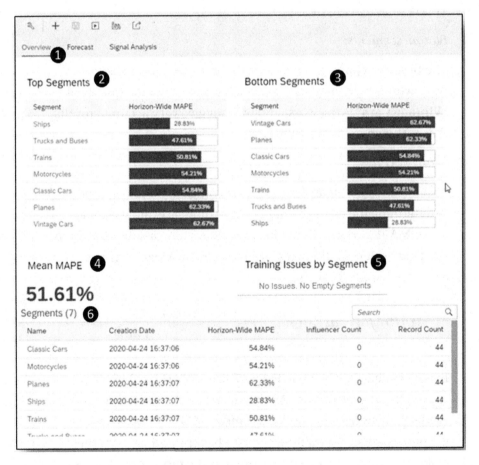

Figure 7-8. *Time Series Predictive Model: Overview*

Overview Section

This section shows the overall outcome of the selected predictive model based on the input dataset.

1. *Section Header:*

   The Header section shows the component under review. In this example we are viewing the Overview section as shown in Figure 7-8.

2. *Top segments:*

   The Top Segments component shows the number of models created and the quality of the models. The topmost model, in our particular case **"Ships,"** has the highest quality, which implies that this model will show the best results with the lowest error rate. This is shown in Figure 7-8.

3. *Bottom segments:*

   The Bottom Segments are similar to the Top Segments but list the lowest quality models first, which implies that the quality of **"Vintage Cars"** is the worst and will not give very accurate results. This is shown in Figure 7-8.

4. *Mean MAPE:*

   MAPE stands for Mean Absolute Percentage Error and is used to determine the average quality of the models evaluated. The Mean MAPE variable shows how much the forecast deviates from the signal value. Mathematically, the value of Mean MAPE is calculated as the following:

$$MAPE(forecast, signal, h) = \frac{1}{N} \sum_{i=1}^{N} \left| \frac{(forecast_{i+h}^{h} - signal_{i+h})}{signal_{i+h}} \right|$$

   In our example, as shown in Figure 7-8, the MAPE is about 51%. Ideally, the lower the mean MAPE, the better the validity of the model that has been created. If the mean MAPE is not as desired, the process of model training needs to be repeated with another set of data to ensure the final model predicts the most correct results.

5. *Training Issues by Segment:*

If there are any issues that will prevent the model from being trained, they are listed here with the reason as to why the training failed. In our case, there have not been any issues as shown in Figure 7-8.

Training a model fails primarily due to issues in the data. The dataset has to be clean without any issues and uniform in typecasting. A dataset of good quality will ensure the model is able to perform better in terms of quality of output provided. Data quality of the dataset should be observed along an ample number of rows for the predictive model to perform and train well. Along with data quality, business knowledge is essential to select the right data for training the model. Based on the type of predictive scenario and the desired outcomes, training data should be curated to deliver the best results.

6. *Segments:*

This component details out further information about the segments created like the records considered and the model creation date and time. To get a better quality of forecasts, SAC's smart predict engine creates multiple models, calculates the individual MAPE for each of the segments over the respective horizon, and averages it out. This average is called a horizon wide MAPE. Typically, a zero value for the horizon wide MAPE indicates a perfect model and as the value of this variable increases, the accuracy of the model keeps falling. This is shown in Figure 7-8.

We have learned about the Overview section, which provides a brief on the overall confidence level of the model. Let us now learn about the components of the Forecast Section.

Forecast Section

The Forecast Section holds the information about the actual forecast for the target variable. Analyzing the forecast section enables further information about how the forecast can be comprehended and further tuned.

Figure 7-9. *Time Series Predictive Model: Forecast*

1. *Section Header:*

   The header component shows the section of the Debrief in section. In this particular case, this is Forecast. The Forecast section shows the actual forecast for the target variable, which is sales by segments. This is shown in Figure 7-9.

2. *Categories:*

   The categories drop-down allows for the forecasts to be segmented by individual items within the Product Category. In this example, we have selected **"Classic Cars"** as shown in Figure 7-9.

3. *Global Performance Indicators:*

   This component in Figure 7-9 shows the quality of forecast for the model in terms of categories. In this example, the Indicator is Horizon Wise MAPE.

4. *Forecast Vs Actual:*

   This chart in Figure 7-9 shows the actual Time Series wherein the forecast is displayed with the error range. A better forecast can be generated with the quality of data used for training the model. Also, a higher volume of training data would ensure a better forecast quality. This subsection shows the target variable values over the period of time selected.

5. *Signal Outliers:*

   An outlier is wherein the predicted value lies far outside the forecast range. Outliers require special attention since their values cannot be explained by the current forecast curve. In our particular example, there is one value that is far out of the forecasted range. This value can be investigated further for data discrepancies of any other issues that might have shown a drastic difference between the actual and forecasted value. This is shown in Figure 7-9.

6. *Signal Anomalies:*

    An anomaly is usually a missing value in the dataset due to which there could be an error in the forecasted value. In our example, there are no anomalies as shown in Figure 7-9.

We have thus learned the forecast section of the predictive model Debrief. Let us now learn about the last section, which provides further information in terms of the Signal Analysis.

Signal Analysis Section

The Signal Analysis section enables in-depth analysis of the Target or Signal Variable that has been selected. The Signal Analysis section is as shown in Figure 7-10.

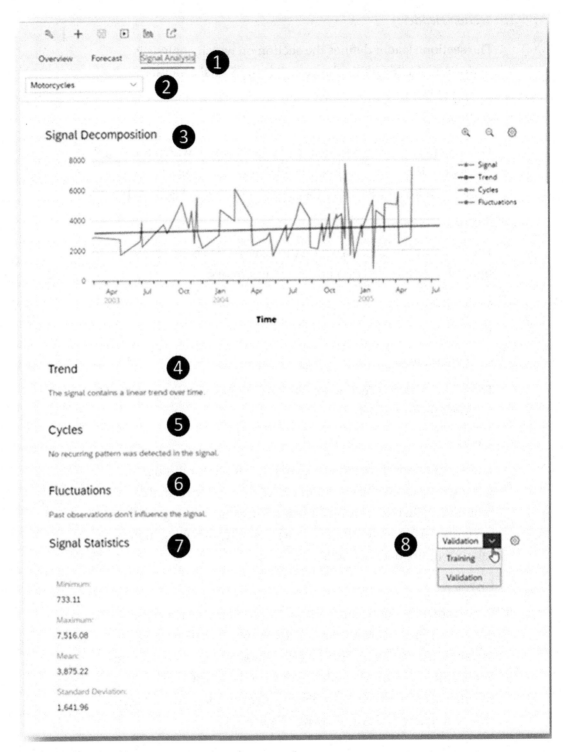

Figure 7-10. *Time Series Predictive Mode: Signal Analysis*

1. *Section Header:*

   The section Header defines the section on which analysis is performed. In this particular case, it is the Signal Analysis as shown in Figure 7-10.

2. *Category:*

   This component allows selection of categories on which the signal analysis is to be done. As per our selection, the forecast is created as per each item in **Product Category** as shown in Figure 7-10.

3. *Signal Decomposition:*

   The signal or the target is where the forecast has been created. The signal is decomposed into further components:

   - Trends

   - Cycles

   - Fluctuations

   The Signal Analysis section allows each of these components to be explored further in detail.

4. *Trend:*

   This component shows how the overall trend is for the signal. A trend is basically the direction where the average signal is progressing. In our example of **Sales**, the trend is rising as shown in Figure 7-10, which brings up the possibility of having higher sales for future dates as well.

5. *Cycles:*

   Cycles are periodicity patterns that help determine if a certain pattern is repeated over time. Cycles enable in determining if certain factors influence the target variable in repetitive patterns. For example, motorcycle sales may see a rise every year in the summer season. In our example, the algorithms have not been able to find any cycles as shown in Figure 7-10.

6. *Fluctuations:*

   Fluctuations are anomalies in the trend but over time. These
   fluctuations also enable end users to determine if certain
   fluctuations have been proving detrimental to the trend of
   the signal variable. In our example, there have not been any
   fluctuations to the trend shown in Figure 7-10.

7. *Signal Statistics:*

   Signal Statistics provide further information on the signal
   including the minimum, maximum, mean, and standard
   deviation. Each of these statistics can be evaluated how much
   further the signal is from the mean value. This is shown in
   Figure 7-10.

8. *Training/Validation Selector:*

   This option enables selecting the option to train the model or to
   validate the outcome of the trained model.

   We have learned in the above section the options available for
   predictive scenarios and how they can be tuned to deliver the best
   possible outcomes for business queries.

In this section, we have discussed Predictive Scenarios available in SAC that enable
end users to effectively use predictive analytics capability to empower data-driven
decision-making. We also learned the step-by-step process to facilitate a Time Series
Predictive Scenario and then trained and debriefed the Time Series model created.

Summary

And this brings us to the end of this chapter, **"Capitalize on Predictive Analytics
through SAC."** We have learned about SAC Smart Predict, a component of SAC's
"Smart Assist" and how it enables rapid predictive models for end users, enabling an
organization to graduate to a Data-Driven Decision-Making culture. We have learned
about predictive scenarios, and the different types of predictive scenarios available
within SAC. Further, we also learned about creating a predictive model and the options
available within SAC that can be used to create a predictive forecast on historical data.

In the next chapter, we will learn how SAC provides a simple but highly effective interface for building custom applications with the SAC Landscape. We also will learn how the custom applications are different from the Stories and Dashboards in SAC. Finally, we will discuss learn the step-by-step process to build a custom application.

CHAPTER 8

Craft Special Business Requirements on SAC via Custom Analytics Application Design

As a global organization, ABC Inc. has multiple teams working on diverse datasets with specific requirements. Each of these teams have, in collaboration with IT teams, built customized versions of analytical applications that are neither standardly offered from the product nor qualified as a standard across ABC Inc. While grappling with the lack of **"Single Value of Truth"** issues, ABC Inc. has recognized that in addition to data challenges, another challenge would be due to the custom applications, be it for the IT teams or business processes and people across the organization, and therefore is an impediment to their business growth.

In addition, ABC Inc. has a huge constraint with the current analytics landscape to customize reports and dashboards as per their requirements and facilitate standardization across the enterprise. Due to the lack of a modern analytical platform, even a minor customization requires programmers to completely rebuild the entire report. Applications that are built to cater to specific customer needs have to be programmatically developed, taking up considerable development effort and later operational maintenance.

The challenges mentioned above are a cause of concern due to their significant operational costs and loss of productivity. ABC Inc. therefore requires that the new and modern analytic platform has to substantially reduce operational and debugging efforts alongside enabling customization that is better or on par with their current analytics

© Vinayak Gole, Shreekant Shiralkar 2020
V. Gole and S. Shiralkar, *Empower Decision Makers with SAP Analytics Cloud*,
https://doi.org/10.1007/978-1-4842-6097-5_8

landscape. Further, ABC Inc. appreciates the negative impact of having multiple solutions in their landscape; and therefore, it needs an integrated custom application development feature that will work in tandem with the standard capabilities of self-service analytics.

Lack of Reusable Components

Significant efforts have been invested at ABC Inc. in development of analytical applications including a few of the strategic dashboards, as also in Learning and Training of its end users. The executives at ABC Inc., just like any other organization, depend on the analytical dashboards to make critical business decisions. It's well established that boardroom, senior executives focus on critical and actionable insights offered on a standard and from the same form, rather than a varying and incongruent form that distracts due to learning time and effort. In view of these aspects, it is essential that ABC Inc. has a custom application development platform that enables high reuse from their objects, and developers should be able to pick components from existing applications to build new ones. The dashboards should be standard in terms of aesthetics and KPIs to enable the executives to rapidly interpret the data and make decisions apart from being intuitively promoting self-service.

Further, ABC Inc. is keen to integrate their Decision Support System with their Enterprise Applications like ERP, thereby enabling tight coupling between action based on insight. For example, the developers can build a dashboard and integrate it in the application that derives actions from insights. Consider a list of expenses that need to be approved or declined. Instead of going from the dashboard into the application, the approver can approve from the dashboard and write back data into the transactional system or ERP. Such a facility is currently unavailable and not easy with the existing Analytics System Landscape. Lack of the above capabilities has left end users at ABC Inc. severely lacking faith in the current decision support system because of their existing analytical solutions and their flexibility.

All of the above business scenarios demand deployment of an analytics solution that should enable rapid self-service supported by augmented analytics while allowing *custom applications deployed with equal speed and alacrity.*

Designing and building an analytic application would not only reduce the manual effort of end users in day-to-day activities drastically but also enable automation and reduction in data the footprint in data collection, transformation, and storage. ABC Inc. can ensure that enterprise standards are maintained across dashboards and applications as well as across the self-service stories built by end users.

The new analytics landscape should be able to provide the following:

1. Customized application development

2. Standardized content development

3. Access to modern analytics tools and language support

4. Offer low TCO

Alignment to Specific SAP Analytics Cloud Capability

SAC enables multiple analytic applications on a single platform. Enabling self-service data analysis as well as planning over a single platform, SAC also enables building custom analytic applications. The custom analytic applications cater to specific requirements while relying on complex scripting to deliver a best-of-the-class experience to end users. Allowing customizations from data connectivity and discovery, as well as user experience, while continuing to easily integrate with the available functionality like BI and planning, SAC delivers a complete package for end-to-end analytics. Let us learn the difference between multiple applications available in SAC before delving deeper to understand how to build custom analytic applications using the SAC Analytics Designer.

Table 8-1. *Comparison of SAC Applications*

Parameter/Analytic application	Standard Report	Story	Dashboard	Digital Boardroom	Custom Applications
Type	Pre formatted reports built for standard data delivery in the form of reports or charts	Data discovery mechanism facilitating answers to specific business queries	Holistic overview of KPIs and specific metrics arranged in pre formatted pages	Collection of snippets from Stories and reports	Custom application built with a pre defined guided path to follow across multiple pages
Purpose	Reporting	Self Service	Performance analysis	Real time answers	Multi purpose
Typical Format	Tabular	Tables, Charts Graphs	Tables, Charts Graphs	Tables, Charts Graphs	Tables, Charts Graphs Custom components
Dependence on IT	Medium	Low	Medium	Low	High
Time to Deliver	Low	Medium	Medium	Low	High
Customization	Low	Medium	Medium	Low	High
Typical consumers	Managers, End Users	Analysts, End Users	Managers, CXOs	CXOs	All

As shown in Table 8-1, standard reports are the easiest to build but offer the lowest level of customization. On the contrary, the highest customization is offered by Custom Analytic Applications, which can cater to each requirement with advanced scripting and programmatic techniques.

Some of the benefits of having a custom analytic application in addition to the self-service Business Intelligence applications are the following:

1. **Flexibility:**

The functionality of self-service reports and dashboards stop at a point where custom capabilities have to be built into the application. This would mean employing a programming language to build the new functionality. However, this would imply sacrificing the advantages of having an integrated platform for analytics. This would also imply the loss of a robust security model inherent to enterprise applications.

Additionally, one of the primary aspects of modern enterprise application is the ability to integrate with other systems. Embedding analytical reports into other applications is one such functionality that can be termed as necessary. For example, internal websites that publish multiple reports and data points can easily be embedded with data from custom applications.

SAC ensures both these expectations are met in terms of building custom analytic applications as well and integrating them into other applications. SAC Analytic Applications ensure flexibility in terms of building applications where the standard **Reports** and **Stories** fail. With simple JavaScript-based scripting, the SAP Analytics Designer is able to fulfill most of the custom features to be built into the application. It also enables multiple components and widgets that can be placed onto the canvas and scripted as per requirements. Integrating into the existing architecture, the Analytic Applications display utmost flexibility in terms of features.

2. **Reusability:**

The standard reports and dashboards usually have limited capability in terms of component reuse. For example, a chart developed in a report cannot be reused as a component in a story or vice versa. Lack of reusability can be time consuming for developers as well as end users and result in redundancy of work.

The components or widgets built within the canvas layout can be published as reusable components for use within the entire enterprise SAC landscape. SAC Analytic Designer also allows for the development of custom themes that can also be published for consumption.

A considerable amount of time can be saved by reusing components. The SAC Analytics Designer also enables creating a new custom color palette that can be reused across other reports, stories, and applications. Custom CSS can also be included to enable further reusability within the SAC Application. This is especially useful when the organization has defined a set of colors for its brands as well as for the enterprise.

Development efforts can be reduced while increasing the efficiency of delivery of the applications to the end users by enabling reusability across the components and widgets.

3. **Insights to action:**

Actionable insights have been known to offer invaluable information to decision-makers to capitalize on. Data-driven decisions enable actions that are more informed and accurate. The starting point of any action has always been a decision albeit powered by data.

Now imagine a situation wherein the starting point is data or information. The end user who is analyzing the dashboard for a set of crucial KPIs discovers very insightful information. The typical process would be to go to the actionable application, for example, the ERP system to complete the action. However, with the SAC Analytic Application, the action can be programmed into the application itself. The end user would not have to leave the data analysis screen but would be able to execute actions from that very screen.

Such a landscape would be multifold beneficial since it would enable actions to be driven by data from within the data application itself. The end user would be able to trigger actions right from the dashboard itself.

SAC enables programmatic flexibility into the traditional information dashboard by embedding data actions into the dashboard. SAC thus enables Insights to Action within the landscape itself.

This feature is especially useful for maintaining consistency and standards and enables a complete 360-degree execution for data actions.

SAC's Analytics Designer offers a custom development environment that previously was available only with Lumira Designer and SAP Design Studio. Offering a complete custom development package, the Analytics Designer enables building applications to be built using connections to data sources or through file uploads. The Analytics Designer can be integrated with other components of the SAC like Planning, Stories, and Connections and as well as other web applications.

The Analytics Designer also allows writing back to source to enable a closed loop functionality that can enable actions to immediately follow decisions. For example, upon viewing the inventory dashboard application, the user realizes that stock needs to be replenished. The user can immediately trigger an action on the source ERP system to put in an order for the particular stock.

Another example would be of rapid approvals from the dashboard itself without the need to switch applications. The approved data would be written back to the transactional system or ERP.

SAC's Analytics Designer thus enables to-the-point-custom application development with a rich variety of scripting techniques enabling ABC Inc. to fill the gap of fulfilling certain requirements not available out of the box. Reusability of components enables developers to rapidly deploy applications without having to build from scratch. The reusability also enables ease of maintenance. Consider there has to be a change in the logic in which Revenue is calculated. The component that calculates Revenue can be changed, thus reflecting the change across all the analytic applications that use the component.

Additionally, since the SAC Analytics Designer is part of the SAC landscape, there is no need to develop additional security but can integrate into the already available security framework.

In this section, we shall learn about the components of SAC's Analytics Designer and then follow on with the step-by-step process of adding components to the designer to build a custom dashboard in the next sections.

We will begin by designing the analytic application and then learn the step-by-step process of adding elements to the application. We will then learn how to add the logic in terms of scripts to components added before finally creating the custom dashboard. We will then debug the application to check for any issues and finally learn the step-by-step process to run the analytic application created.

Note   Since SAC is a constantly evolving product with SAP introducing new features every quarter, some of the actual screens might be different from what are shown in the book.

SAC – Analytics Designer

Learning to build a custom application in the next section with a step-by-step process requires us to know features and functionalities of the SAP Analytics Designer in detail. You can refer to Appendix B for the entire screen and various settings and their implications and various components of the SAC Analytics Designer. It also discusses how the components help in building a custom analytics application and would facilitate appreciating the step-by-step process to create an analytic application in the following section.

Creating an Analytic Application

In this section, let us build an application that would display the sales by country on a click of the corresponding button and display the sales by product on a display of the button corresponding to this option. At one point of time, only one chart or graph should be visible and the other should not be rendered on the canvas. This application can be used to view the **"Sales by Product"** or by **"Sales by Country"** on the same available screen size.

1. **Designing the Analytics Application:**

Before any application can be built, it should be designed as per the requirements. For building a mock report, a wireframe built to support the requirements would enable end users to get a view of the application before it can be completely built. Let us build a mock-up application as per the requirements. The mock-up application looks like Figure 8-1. As can be seen, on a click of button 1, the sales by country would be displayed as a bar and on a click of button 2, the sales by product would be displayed as a pie chart.

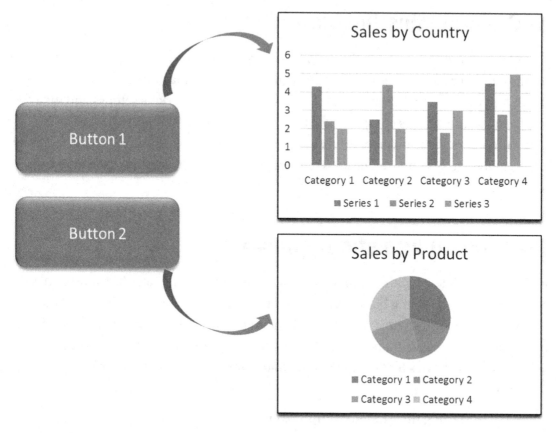

Figure 8-1. *Design for Sample SAC Analytic Application*

In this subsection, we have created a mock-up with the logic for implementation. In the next subsection, let us learn the step-by-step process for adding the first component, which is the bar graph to the SAC Analytic Designer Canvas.

2. **Adding a Graph to the Canvas:**

Once the mock-up for the analytic application has been created, let us learn the step-by-step process to input a graph to the canvas. The process is shown in Figure 8-2.

Figure 8-2. *Step-by-Step Process to Add Graph to Canvas*

Step 1: From the **Main Menu**, click on **"Create"** and then on **"Analytic Application."** The **Analytics Designer** comes up as shown in Figure 8-2.

Step 2: From the **"Insert"** option in the top menu, click on the ≣ Symbol as shown in the figure.

Step 3: This brings up the **Model Selector** Window. Select the model. Let us select the **"Demo_Sales_Sample"** model for creating the graph as shown in the figure. Models that have already been created can be used as well as custom data sources that can also be created for creating complex applications.

Step 4: Once the Model is selected, an empty graph comes up in the canvas. Use the Designer Panel to select the measure **"Sales"** and dimension **"Country"** as shown in Figure 8-2.

Step 5: The bar graph now shows the data for Sales as per Country. This bar graph will correspond to the bar graph we have shown in Figure 8-1 wherein we have designed the mock-up application. On clicking the first button, this bar graph should be visible.

Step 6: Name the bar graph **"Bar_SalesByCountry"**. This can be done from the outline panel on the left of the canvas as shown in Figure 8-1.

Step 7: From the styling menu on the right, uncheck **"Show this item at view time"** as shown in Figure 8-3. This will ensure that the bar graph does not load when the application is run. It should be visible only by clicking Button 1.

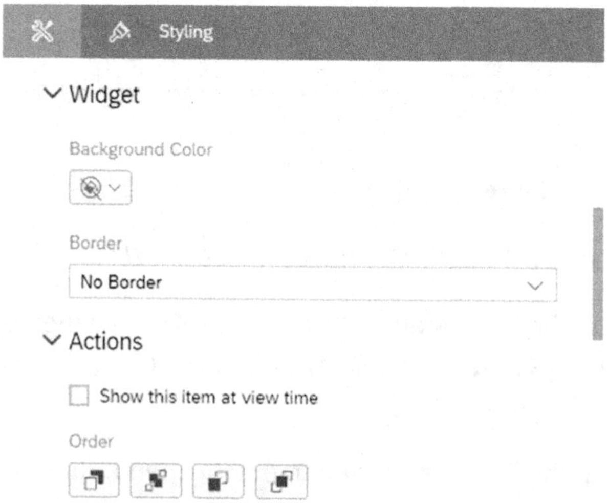

Figure 8-3. *Graph Styling*

In this subsection, we have learned the step-by-step process for adding a bar graph to the analytic application. In the next subsection, we will learn the step-by-step process to add the next component, which is a donut chart to the canvas.

3. **Adding a Chart to the Canvas:**

We have seen in the previous section how a bar graph can be added to the analytics application canvas for creating the first component of our application. Let us now follow a similar step-by-step process to add a chart to the canvas.

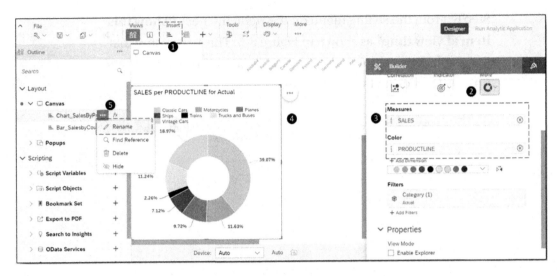

Figure 8-4. *Step-by-Step Process to Add Chart to Canvas*

Step 1: From the **"Insert"** option in the top menu, click on the ⊫ Symbol as shown in Figure 8-4.

Step 2: An empty chart comes up. The chart would be based on the same model selected in subsection 2, Step 3, which is **"Demo_Sales_Sample"**. We would be building a pie chart as per our specifications defined in subsection 1 and hence let us select the **Donut Chart**, which is a variation of the Pie Chart, from the **Builder Menu** as shown in Figure 8-4.

Step 3: Select the dimensions and measures from the **Builder Menu**. Since we have to display the Sales by Product, let us select **"Sales"** for the measure and the **"Product Line"** for the dimension as shown in Figure 8-4.

Step 4: The Donut Chart for Sales values by Product comes up as shown in Figure 8-4.

Step 5: Let us now name the Donut Chart created. Since it will depict the Sales by products, let us name the chart **"Chart_SalesByProduct"**.

Step 6: From the styling menu on the right, uncheck **"Show this item at view time"** as shown in Figure 8-5. This will ensure that the graph does not load when the application is run and will come up only by clicking Button 2.

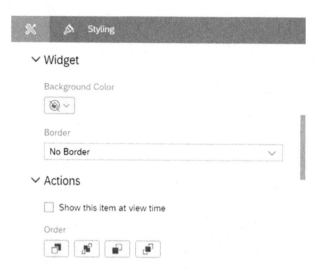

Figure 8-5. *Chart Styling*

In this subsection, we learned the step-by-step process to add the donut chart to the canvas. We have now added two components to the canvas, a bar graph and a donut chart. In the next subsection, we learn the step-by-step process to add two buttons to the analytic application canvas.

4. **Adding Buttons to the canvas:**

We have now added the two primary components to the application. Now let us add the buttons. The button widgets are placeholders while the actual script will be attached to them to display the desired effect on clicking on the buttons. The script will hold the actual logic on the action that will execute when the buttons are pressed.

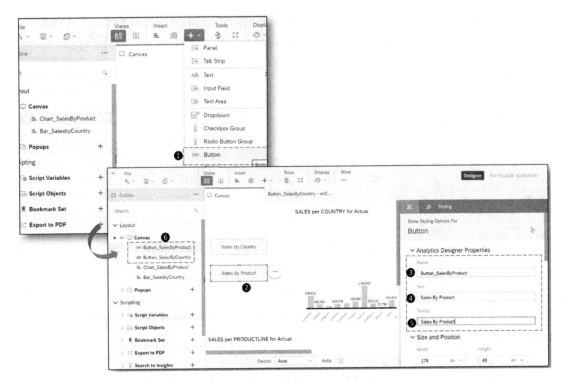

Figure 8-6. *Step-by-Step Process to Add Button 1 to Canvas*

Step 1: From the **"Insert"** option in the top menu, select the option to add a **"Button"** by clicking on the '+' symbol as shown in Figure 8-6.

Step 2: The **Button** is added to the Canvas as shown in the figure.

Step 3: From the **Button Styling Panel**, give an appropriate name to the **Button**. Let us name the Button **"Button_SalesByProduct".**

Step 4: The **Text Component** stands for the text to be displayed on the Button when it is actually rendered. Let us name the button **"Sales By Product."**

Step 5: The tooltip offers mini help about the button. Any additional helpful information can be displayed in the tooltip. Let us keep it simple and in the tooltip, let us put **"Sales By Product"** to maintain consistency across the application.

We have learned to add the first button to the analytics application as per the step-by-step process above. Let us now learn how to add the second button to the application.

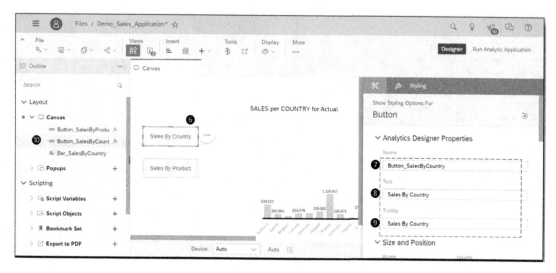

Figure 8-7. *Step-by-Step Process to Add Button 2 to the Canvas*

> Step 1: From the **"Insert"** option in the top menu, select the option to add a **"Button"** by clicking on the '+' symbol as shown in Figure 8-7.
>
> Step 2: The second **Button** is added to the Canvas as in Figure 8-7.
>
> Step 3: From the **Button Styling Panel**, give a name to the button. Name the button **"Button_SalesByCountry".**
>
> Step 4: The Text Component stands for the text to be displayed on the Button when it is actually rendered. Name the button **"Sales By Country".**
>
> Step 5: In the tooltip, let us put **"Sales By Country"** to maintain consistency across the application. This is shown in Figure 8-7.

In this subsection we have learned the step-by-step process for adding two button components to the analytic application canvas. We have now successfully added a bar chart, a donut chart, and two buttons to the canvas. In the next subsection, let us learn the step-by-step process to aesthetically place the components added above onto the analytical application canvas.

5.   Placing components on the canvas:

Once the components have all been created, the Application Screen needs to be created to respond to the clicks on the buttons. The components should be placed aesthetically to enable the best user experience. Place the buttons on the extreme left and the charts one above the other as shown in Figure 8-8.

This is to ensure that both the graph and the chart appear at the same location when the corresponding button is pressed.

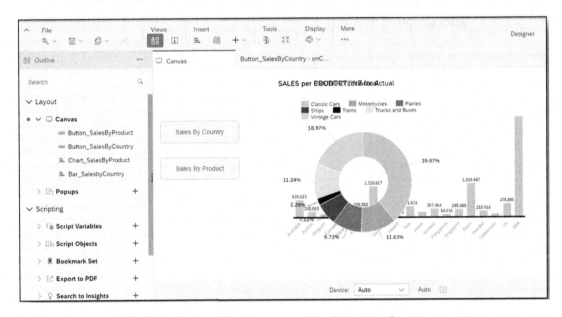

Figure 8-8. *Final component placement on Analytic Application*

In this subsection we have learned the step-by-step process to place components aesthetically so as to follow the design defined in subsection 1 of this section. In the next subsection, let us learn the step-by-step process to add the logic in the form of scripts to the buttons as per the design.

6.   Adding Scripts to Buttons:

We have in the previous subsections from 1 to 5 placed containers and placeholders within the application. Now let us learn the step-by-step process to program the logic within the application. The scripts on the buttons will enable the graph and the chart will be displayed alternatively. Refer to the mock-up created for the application in the subsection **"Designing the Analytics Application."**

Some of the salient features of the scripting functionalities of the SAC Designer are as defined below:

- The scripting language in Analytics Designer is TypeScript, which is a limited subset of JavaScript.

- Content assistance is enabled for developers to have an easy-to-use view of the scripting functions available.

- Commenting and syntax check features are available within the SAC Designer scripting window.

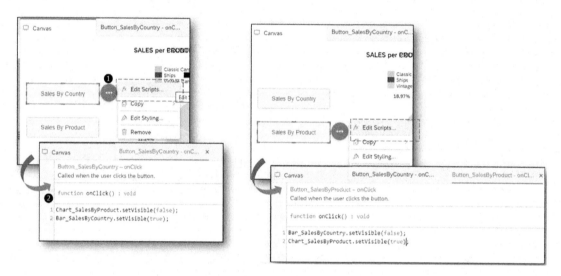

Figure 8-9. *Step-by-Step Process to Add Scripts to Button Components*

Step 1: Click on the **Button Menu** beside the Button component as shown in Figure 8-9. Click on **"Edit Scripts."** This brings up the Scripting Application for the button as shown in the figure.

Step 2: In the Scripting Application, create the script for implementing the logic. On clicking the Button for Sales by Country, the Sales by Country Graph should appear as per the design in section 3.1. Hence, the script should be:

```
Bar_SalesByCountry.setVisible(true);
```

Meanwhile the Donut Chart for the Sales by Product should not be visible. Hence the script should be:

```
Chart_SalesByProduct.setVisible(false);
```

This is as shown in Figure 8-9.

Step 3: Similarly, for the second Button, which is the Sales by Product, Click on the **Button Menu** and Click on **"Edit Scripts."** This is shown in Figure 8-9.

Step 4: In the scripting window, create the below scripts:

```
Bar_SalesByCountry.setVisible(false);
Chart_SalesByProduct.setVisible(true);
```

Step 5: Once the scripts have been created, Save the application.

In this subsection, we have learned to add scripts to the buttons in the analytic application. Since we have now successfully created the application, let us debug and run the application.

In the next subsection, let us learn the process to debug the application before running it.

The complete SAP's Developers Guide for SAP Application Development can be found at `https://www.sapanalytics.cloud/analytics-designer-handbook/`

7. **Debugging the Analytic Application:**

Debugging enables application errors to be resolved. The Debug panel in SAC Designer enables developers to resolve errors encountered during developing the application.

In case of any issues in the application, the errors come up in the Info Panel as shown in Figure 8-10.

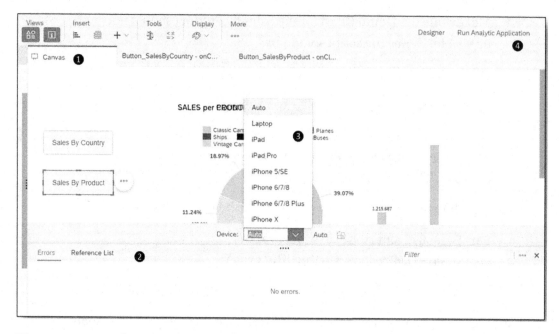

Figure 8-10. *Debug Analytic Application*

Figure 8-10 shows the finalized application with the placement of the widgets and components across the canvas. The components of this screen are the following:

1. **Canvas:**

   This is the container that holds all the components of the analytic application. The application would also run on the type of canvas selected. This is shown in Figure 8-10.

2. **Info Panel:**

   The info panel shows any errors encountered during creating the application. If there are any references, they are shown as well in Figure 8-10. This panel is especially important during debugging.

3. **Device Simulation:**

   Simulation enables testing the application as per the screen size of the primary device where the application will be deployed. To test and deploy the application to a particular device, the simulation panel comes in very handy, which shows the most popular devices.

Select **"Auto"** to enable automated placement of the components on the responsive canvas, which will adjust as per the screen size. This is shown in Figure 8-4.

4. **Run Analytic Application:**

   The **"Run Analytics Application"** enables running the application as per the device selected in the Device Simulation option. We will learn the step-by-step process to run the application in the next subsection, number 8.

In this subsection, we have learned the options available in the SAC Analytics Designer to debug the application. In the next subsection, let us learn the step-by-step process to run the analytic application we have created.

8. **Running the Analytic Application:**

The **"Run Analytic Application"** option from the top menu enables the application to be run. We have learned about the Top menu in detail in the section "SAC – Analytics Designer."

The completed Analytic Application is as shown in Figure 8-11.

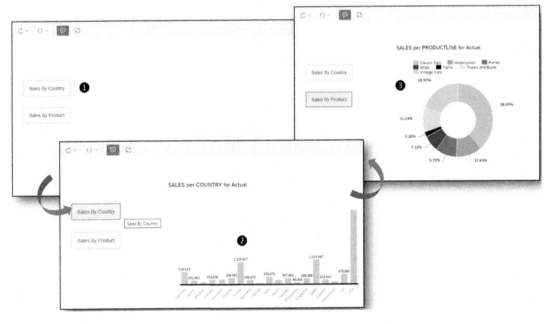

Figure 8-11. *Step-by-Step Process to Run Analytic Application*

Step 1: The Analytic Application is shown in Figure 8-11. The initial screen does not have any graphs or charts but only two buttons **"Sales by Country"** and **"Sales by Product".**

Step 2: Click on the Button **Sales By Country** and the Bar Graph for **Sales by Country** comes up. Notice that the Donut chart for the **Sales by Product** is not visible. This is shown in Figure 8-11.

Step 3: Clicking on the button **Sales by Product** brings up the Chart for **Sales by Product**. The Bar graph for **Sales by Country** is no longer visible as shown in Figure 8-11.

In this section we have learned the step-by-step process to run the analytic application we have created. In the next subsection, let us learn the step-by-step process to share the completed analytic application to other end users.

9. **Sharing the Analytic Application:**

The completed Analytic Application can be shared with specific users by using the **"Share Application"** option. Sharing an application enables the developer to share with other users and developers. This is shown in Figure 8-12.

Figure 8-12. *Share Analytic Application*

Step 1: Click on **"Share Application"** from the **Top Menu**. This brings up the **Share Application** Window. Select the **"Users or Teams"** to be added for access to the application. This is shown in Figure 8-12.

Step 2: Select the level of access as shown in the figure.

Step 3: The users with whom the application has already shared is shown in Figure 8-12. Click **"OK"** to share the Application with the users.

We have seen the step-by-step process for creating a custom analytic application and running it on the screen. This functionality enables organizations to build functionality that is not available in the standard story builder functionality.

We have now learned the step-by-step process to create and run a custom analytic application with the SAC Application Designer.

SAC enables programmatic flexibility into the traditional information dashboard by embedding data actions into the dashboard. SAC thus enables Insights to Action within the landscape itself.

This feature is especially useful for maintaining consistency and standards and enable a complete 360-degree execution for data actions.

Summary

And this brings us to the end of this chapter on how SAC supports building custom applications for developing specific business scenarios and integrating with standard use cases for self-service analytics. In this chapter we have learned the following:

1. The importance of having the ability to build custom applications to cater to specific scenarios.

2. Designing, creating, debugging, and then running a custom application through the step-by-step process using SAC's Analytics Designer.

3. Customer benefits of an SAC custom-developed analytic application.

In the next chapter, we will learn how SAC facilitates a robust and secure landscape with customizable security at each level of the application.

CHAPTER 9

Design A Secure Platform Using SAC

ABC Inc. is a large, complex, and global organization and so far has managed to ensure that all of its IT assets are securely protected under multiple layers. With the rapid rise of internet technologies and its acceptance as one of the pillars of modern infrastructure, cybersecurity incidents have witnessed an exponential growth. Security concerns have been growing over the last decade, which forced ABC Inc. to review every possible aspect of its information security landscape. Data and insights from data are strategic levers for surviving and competing in the current environment, and ABC Inc. recognizes the paramount importance of securing its analytics landscape.

Further, all of ABC Inc.'s IT systems have been on-premise complying to all enterprise security standards. Keen to leverage benefits from cloud technology, ABC Inc. knows that to build an analytics landscape on the cloud, it is important that the new landscape should secure and follow industry security standards. Also, as ABC Inc. moves to multiple cloud-based solutions, security plays a primary role in the design of the entire enterprise architecture. The digital transformation would see end users access data not only from desktop PC's but from multiple other devices, including mobile devices.

In addition to the security framework laid down by ABC Inc., the applications within the enterprise landscape are also expected to meet world-class security standards. These applications are required to be robust and yet easy, that is, tightly sealed from any external and unauthorized access at the same time, enabling a platform for easy collaboration within internal stakeholders and users.

Some of the other requirements that ABC Inc. expects from the new landscape is with the enabling of collaboration within the organization. However, collaboration should be bound by robust security concepts and multilayered security. As with a typical organization, ABC Inc. has multiple divisions with layered hierarchies and access levels,

© Vinayak Gole, Shreekant Shiralkar 2020
V. Gole and S. Shiralkar, *Empower Decision Makers with SAP Analytics Cloud*,
https://doi.org/10.1007/978-1-4842-6097-5_9

which are applicable to certain roles at the hierarchy levels. Data-level security hence plays a pivotal role, which has to be augmented by a robust model of roles that defines which functionality to enable for the end user.

ABC Inc.'s current security needs for the new system thus encapsulate two levels: authentication and authorization.

1. **Authentication:**

    Authentication is the process of enabling an end user to log in to the system. An unauthorized user should be prevented from accessing the landscape. This can be achieved by using a username and password or through enterprise SSO. This is the first step to ensure a robust security model. The built-in authentication of an application is of utmost importance to ensure that there is no entry into the system from unauthorized users.

2. **Authorization:**

    Authorization is the second stage of access. Once authenticated, the end user should be able to access only the aspects of the landscape for which access has been granted. Authorization is enabled through specific rights and roles that enable each end user to perform specific activities without hinderance.

A combination of authentication and authorization should be used to secure end-user access. Security is a complex but very important component of enterprise architecture. A landscape with weak security could expose sensitive data to unscrupulous elements. The current analytics landscape at ABC Inc. has a robust authentication but lacks authorization concepts. In the new landscape, the requirement is to have a secure analytics landscape that can be accessed and can also be integrated using a single sign-on in the future for enterprise-level security.

The new analytics landscape at ABC Inc. should thus incorporate best-in-class authentication and integrate authorizations with other applications while maintaining security within the analytics landscape. For a cloud-based landscape, the cloud data center should adhere to the latest global as well as regional policies for security.

Alignment to Specific SAC Capability

SAC enables a robust security system and system administration capabilities. Both the security concepts of authentication and authorization are well taken care of by SAC. Access to the SAC landscape can only be granted by the Administrator and can be accessed only via the URL provided. From the authorization perspective, SAC allows creation of Users, Teams, and assigning Roles to them to allow only certain capabilities available to them. SAC also has a folder structure that enables creation of private and public versions of objects. Each of these objects can be further assigned security by allowing only certain users or teams to access the same.

Information security is an intricate process and therefore let us first understand the components of the security architecture then learn how security models are created to ensure a secure environment for the SAC application. We will first learn to create a User and then include the User into a Team. We will then learn to create Roles with the right permissions and assign it to the User. Finally, we will learn how to monitor data changes to ensure only authorized data is imported into the Models.

User

A user is a unique entity with the landscape of any system. In SAC, each user stands for the individual who can log in to the system. The user is entitled to perform certain tasks within the landscape as per Permissions assigned to the user.

The representational User concept is shown in Figure 9-1.

User

Figure 9-1. *User*

Teams

A User can be part of multiple Teams. Team is a concept unique to SAC and enables aggregating users into business functions or groups. This enables Users to be managed efficiently across multiple roles, functions, and folders. Teams also enable efficient maintenance of the security landscape.

The representational Team is shown in Figure 9-2.

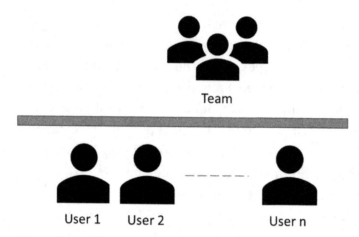

Figure 9-2. Team

Permissions

Permissions allow users to perform certain tasks within the SAC landscape. They also define what level of access a User has within the SAC landscape.

Permissions can also be assigned to shared folders to ensure only certain users gain access to the objects within those folders. Multiple permissions can be grouped together to form a role. A role can be assigned to a User or Team. Permissions are the most granular level within the security model.

The representation of a Permission is shown in Figure 9-3.

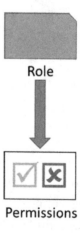

Figure 9-3. Permissions

Roles

A Role is an aggregation of Permissions. A Role not only allows a logical grouping of applicable Permissions but also allows the organization to structure the security model by considering the levels of access for each role as well as for folders within the SAC landscape.

A Role can be assigned to a User or multiple Roles can be assigned to one User. A default Role can also be defined to assign a newly created user with the role.

The relation between Roles and Users is shown in Figure 9-4.

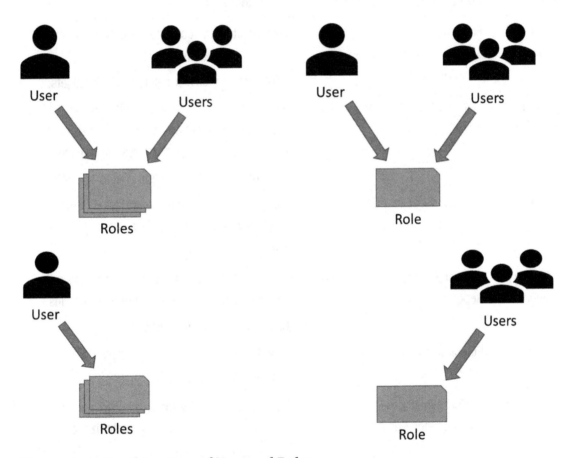

Figure 9-4. *Combinations of User and Roles*

SAC pre-delivers standard Roles that can be utilized for most common tasks. The standard Roles can be managed though the security application. For User Permissions that do not fit into these combinations, custom roles have to be created. The Standard Roles are organized by License Type as per the subscription to SAC. The Business Intelligence License includes the BI Admin, BI Content Creator, BI Content Viewer,

Boardroom Creator, and Boardroom Viewer roles, along with deprecated and custom roles for specialized BI content. The Planning Professional role, which is an add-on to Business Intelligence, includes the Admin and Modeler roles and Planning Standard includes the Planner Reporter, Viewer, and View VDT roles.

The standard Roles provided by SAC are shown in Table 9-1.

Table 9-1. *SAC Standard Roles*

Role	Authorization	Description
System Owner	Full Privileges	Full unrestricted access to the entire system. Only one user can have this role.
		Can create, modify, or delete custom widgets.
Admin	Planning Administrator: Full Privileges	All permissions and task authorizations in the system. Can set up users, roles, system administration, and perform system transports.
		Can create, modify, or delete custom widgets.
Modeler	Planning Modeler: Full Privileges	Includes all task authorizations required for managing models and dimensions.
		Can view analytic applications.
Planner Reporter	Planner Reporter: Planning and Reporting Privileges	Includes all authorizations required for executing planning activities and automated discoveries.
		Can view analytic applications.
Viewer	Planning Viewer: Read Privileges	Includes only authorizations to view the data.
		Can view analytic applications.
BI Admin	Business Intelligence Administrator: Full Privileges	Includes all authorizations except for planning. Has access to predictive task authorizations.
		Can view analytic applications.
BI Content Creator	Business Intelligence Content Creator: Create and Update Privileges	Includes authorizations required to create or modify BI models and dimensions.
		Can view analytic applications.

(continued)

Table 9-1. (*continued*)

Role	Authorization	Description
BI Content Viewer	Business Intelligence Viewer: Read Privileges	Includes read-only privileges to non-planning data. Can view analytic applications.
Application Creator	Application Creator: Analytics Designer Privileges	Includes authorizations required to create or modify analytic applications.
SAPCP Content Creator	SAP Cloud Platform Creator: Create and Update Privileges	Includes all authorizations for non-planning models and dimensions. This privilege allows access to SAP Cloud Platform as a data source.
SAPCP Content Viewer	SAP Cloud Platform Viewer: Read Privileges	Includes authorizations to view non-planning data. This privilege allows access to SAP Cloud Platform as a data source.
Boardroom Creator	Digital Boardroom: Create and update Privileges	Includes authorizations to create, modify, or delete Digital Boardroom Agendas.
Boardroom Viewer	Digital Boardroom: Read Privileges	Includes authorizations to view Digital Boardroom agendas.
Predictive Content Creator	Predictive Scenarios: Create and update Privileges	Includes authorizations to create, modify, delete, and view predictive scenarios.
Predictive Admin	Predictive Scenarios: Full Privileges	Includes authorizations available in SAC. This role is necessary to publish a predictive scenario using the PAi (Predictive Analytics Integrator) to other applications.
Translator	Read Privileges	Includes all authorizations to create, update, delete, and view translations.

All of the above Standard Roles assign Permissions globally and the subsequent access has to be managed manually. As an example, if the Modeler Role is assigned to a user, access to all Models is available to the User by default. The Admin has to then define which Models the user should have access to for editing. Standard Roles can also be edited to configure and refine the permissions assigned to the role.

Custom Roles can be created by either copying a Standard Role as per the License Type or building one from scratch. Custom Roles enable filling the gap for Roles not available under the Standard Roles. This ensures better maintenance and governance across the SAC landscape. Custom Roles provide the much-needed flexibility in building a robust security model by enabling the administrators to enable only certain permissions for specific users. For example, the BI Content Creator role is provided by SAC as a standard role. This role provides the user who is assigned to it the ability to create models and stories. A custom role can be created by copying the BI Content Creator Role and creating a BI Story Creator role, which will provide the user with only the permission to create stories. This can be done by copying the BI content creator role and removing the permission to create and modify models from it and saving the new role as BI Story Creator.

Security Model

A security model is a logical structure for the overall structure that defines how the security has to be defined within the enterprise. A security model lays the foundation within SAC to define creation of roles, users, and appropriate rights. Though there is no component within SAC to define a security model, to define a logical model is a best practice for setting up the correct security.

A sample security Model is shown in Table 9-2.

Table 9-2. *SAC Security Model*

SAC Folder (Root)	SAC Folder Level 1	SAC Folder Level 2	SAC "Teams"	SAC "Role"	Role Access Level
Public	N/A	N/A	N/A	SAC_ADMIN	Full Control
Public	Sales Reporting	Models (View)	Sales_Team	SAC_Sales	Custom
Public	HR Reporting	Models (View)	HR_Team	SAC_HR	Read
Public	Finance Reporting	Models (View)	Finance_Team	SAC_Fin	Read, Execute, Share
Public	Development	Projects	Dev_Team	SAC_DEV	Create, Read, Update, Delete (Custom), Execute, Maintain (Custom)

Public Folder

A Public folder enables organization of SAC components like models and stories into folders. The Public folder is accessible to all users but can be restricted by creating the correct folder structure and assigning rights to the folders. A parent and child folder structure would enable end users to consume rights of the parent folder. This is called Rights Inheritance.

Models and Stories can be shared with other users by placing them in Public Folders or any of the subfolders. Saving a Model or Story into a Public Folder enables all users who have access to the Folder to have access to the Model or Story. For a user to be able to access and run a Story, at least Read Permission should be granted to the user on the Model. A write permission enables the user to create and modify objects such as models, reports, and stories within the folder.

A structure as shown in Figure 9-5 also ensures content organization across the landscape and the right authorization to the content. Coupled with roles, authorizations ensure that not just the right user has access but also the right user has access to the right content.

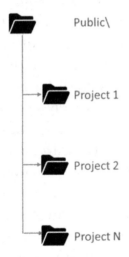

Figure 9-5. *Folder Structure*

Private Folders

In addition to the public folders wherein multiple users have access to content, SAC also has the concept of private folders also known as **"My Files."** These folders ensure that an end user has a space wherein private versions of stories and reports can be maintained. No other user except the Administrators or the System Owner can access this area created for the SAC Users. The **"My Files"** folder also acts as a development space in case of any changes to existing stories. The developer or end user can copy an existing story, modify the story, and use it as per requirements.

We have learned about the basic building blocks of the security in SAP Analytics Cloud. We have learned how SAC enables multiple levels of security across Users and Folders. The security model provides a logical method of building robust security across the landscape.

Let us now learn the benefits of implementing security within SAC and how it can help ABC Inc. build a secure landscape.

1. **Reduced TCO:**

Security is one of the most important aspects for any organization. With cloud applications and interfaces across multiple systems, the security aspect of applications gains all the more importance. A strong security framework ensures that data within the organization is not only secure but also utilized to the fullest. A strong native security also ensures that additional resources are not spent in building a security framework resulting in additional cost in deployment and subsequent maintenance.

SAC's strong security framework ensures users are both rightly authenticated and rightfully authorized to access the appropriate data. Multilevel security in terms of permissions and roles ensures the security is streamlined and each component is rightfully authorized. Users and Teams structure ensures that users are categorized into groups and there is no outlier. SAC security offers multiple features and options that can be combined to create the security model most suited to the needs of ABC Inc. Unauthorized access is restricted by SAC's native authentication mechanism whereas further authorizations ensure the data access is restricted, and the right data is available only to authorized users.

Further security is enabled by the folder structure in SAC. With the ability to create folders and categorize them as per business lines, SAC provides flexibility in saving objects into the appropriate folders, enabling end users to have access only to specific data and content. Unauthorized access is thus restricted from the content level as well with the permissions within SAC.

Additionally, SAC also enables secure sharing of content enabling collaboration across the landscape without any involvement of a third-party tool. Rights can also be requested for within the system and assigned without any additional authorization-requesting mechanism.

All of the above features ensure that ABC Inc. can fully utilize the native security built within SAC and save significant costs and resources in implementing security across the new analytics landscape.

2. **Improved user experience:**

Information technology landscapes have been growing rapidly and along with them data volumes as well. Data analysis provides the edge to organizations in a highly competitive market. With data analysts and citizen data scientists engaged in data mining, the end-user experience plays a very vital role in the enablement of data traversal. Any technology application or landscape is expected to provide the best in terms of engaging the end user and enabling the data analysis process.

SAC provides a secure platform that enables users to view and work with only those folders for which access has been granted. For example, a User with access to the Folder "Sales" will be able to view only that folder. Another user with access to "Finance" will be able to view and work with that folder.

With a focused security model, end users gain the most benefit in terms of navigation around the landscape. Only those folders are visible to the end users for which authorization has been created.

This ensures end users do not spend valuable time navigating the SAC landscape and can quickly land at the destination without much effort. Additionally, SAC also allows end users to request roles, which enables a seamless workflow.

3. **Reduced Risk:**

Cloud applications bring along an additional risk due to security concerns over data centers and access over the internet. ABC Inc. with a vision to have a fully cloud-based enterprise architecture over a period of time has to also cope with the risks that accompany such a migration to the cloud. Additional risk assessments and security audits are time consuming during the initial transition and would be a matter of concern to the plan of ABC Inc. having a fully cloud-native architecture. The SAP Trust Center website lists the stringent global and regional security standards followed by SAP data centers. With a dedicated platform to report security incidents, SAP also ensures that customers who do encounter a security issue can report the issue immediately to the SAP Cloud security team for analysis and rapid resolution.

Additionally, security researchers at the SAP Security Response Team are continually evaluating the vulnerability of the cloud landscape to ensure that the data centers and cloud platform is ready to face the latest security threats.

With a strong security framework and the flexibility to customize as per needs of ABC Inc., SAC enables reduction of risk across the enterprise. ABC Inc. can rapidly scale up to the SAC security framework and design a security model. Built over compact blocks of security objects, the security model can be implemented with minimal risk. SAP enables all the latest security within the data centers, further enabling customers like ABC Inc. to make security audits without having to put in additional time and efforts.

SAC is built natively on the SAP Cloud platform, which adheres to stringent security standards over the Identity Authentication as well as Identity Provisioning. SAP also enables ensuring integrative availability as well as authenticity while the user-level security is applied by the SAC admin at the application level. SAC security also enables adherence to security standards and complies with organizational-level audit standards as well as international standards.

We have learned about the benefits SAC offers in terms of security and how it can help customers like ABC Inc. to rapidly implement and scale up using the building blocks of security. Let us now learn how SAP plans to enhance the security aspects further in the coming quarters as per the latest road map.

In the next section, let us learn the step-by-step process to create security across SAC.

Creating Security in SAC

We have seen in the previous section the components of SAC security and how the components can be used to create a robust security model within the SAC landscape. Let us now learn the step-by-step process to create end-to-end security within SAC. We will learn about how to create Users and then create Teams and frame the security model as discussed in the section **"Alignment to Specific SAP Analytics Cloud Capability."**

SAC provides the option to create security from the main menu. This is shown in Figure 9-6.

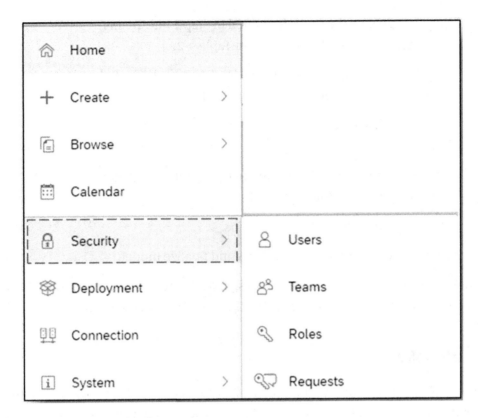

Figure 9-6. *SAC Security Structure*

Security set up in SAC is multilayered and needs to be designed considering multiple factors for levels of access, permissions, hierarchy of roles, and access to folders within the SAC landscape. Let us now explore the step-by-step process to implement each of the steps.

Designing a Security Model

A security model is the output of the design phase for implementing security. As we learned in the previous section, a security model is the framework for setting up a robust security infrastructure. A sample security model is as shown in Table 9-2.

A security model can be created by following the steps below:

> Step 1: Identify the Users who would be accessing SAC. Create the list of Users.

> Step 2: Identify the teams or groups who would be accessing certain folders. Create a list of Teams.

Step 3: Segregate the Users identified in Step 1 into Teams identified in step 2. This is shown in Table 9-2.

Step 4: Map users to standard Roles and create custom Roles in case features do not fit into the standard Roles.

Step 5: Finalize Folder structure as per the projects or Lines of Business.

Step 6: Assign Users or Teams to the Folders. Take special care to factor in the inheritance of Rights.

Step 7: Restrict access to Connections.

Step 8: Review with all stakeholders and finalize model.

In this subsection, we have learned to design a security model. Let us learn the step-by-step process to create users within the SAC landscape in the next section.

Creating Users

Once the model has been finalized for the Users, Roles, and Teams to be created, the corresponding security objects would need to be created in the SAC landscape.

Before we learn the step-by-step process to create the security objects for Users, Roles, and Teams, let us first learn about the components of the Users Management window shown in Figure 9-7, along with a description of the eight options.

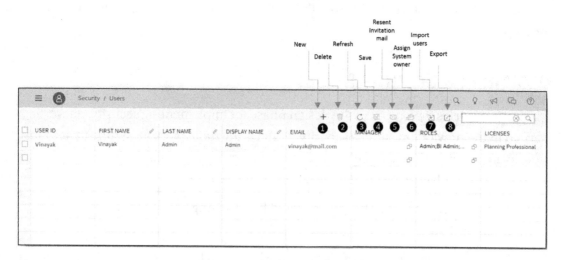

Figure 9-7. *SAC Users Management*

1. **New:**

   This option allows the addition of a new User in the SAC landscape. A User is each individual who can login and execute certain tasks with SAC.

2. **Delete:**

   This option allows for the deletion of an existing User from the SAC landscape. This is especially useful when access to the system has to be revoked due to a change in an organization role or separation from the organization.

3. **Refresh:**

   In certain cases when changes to the security screen do not get reflected, they have to be refreshed manually. The Refresh option allows the manual refresh of the security screen.

4. **Save:**

   The Save option allows saving changes post creation, deletion, or changes to the users in the security screen.

5. **Resend Invitation Email:**

   Once a user is added to SAC, an invitation email with the link to the SAC tenant is sent to the user for the first-time login. In case of issues and the email does not go through, this option can be used to resend the invitation email to the user.

6. **Assign System owner:**

   A System Owner is the overall owner of the SAC tenant. A system owner can delegate rights to another user by clicking on this option. At one point in time, a SAC instance can have only one system user.

7. **Import users:**

   In case multiple users need to be added to the system, this option can be used to import multiple users from a csv file in one go. This option is especially useful while setting up the SAC landscape for the first time when multiple users need to be added in bulk.

8. **Export users:**

As users can be imported in bulk, they can also be exported out of the SAC landscape. This option enables the entire list of users to be exported out of the landscape as a csv file.

Now that we have learned about the components of the user creation screen, let us learn the step-by-step process to create users:

1. **Creating users individually:**

SAC allows creating users by adding information manually or by importing a list of users. While creating a user, the below information has to be entered:

- User ID
- First Name
- Last Name
- Display Name
- Email
- Manager
- Roles
- Licenses

The Entire process for creating a user manually is shown in Figure 9-8.

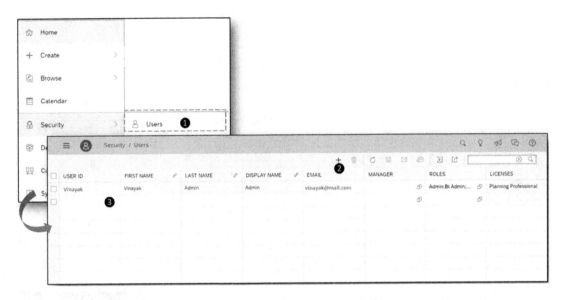

Figure 9-8. *Step-by-Step Process to create users*

Step 1: From the **Main Menu**, go to **"Security"** and click on **"Users"** as shown in Figure 9-8.

Step 2: Click on the '+' symbol to add a new **User** to the environment.

Step 3: Fill in the details of the **User** as shown in Figure 9-8. Once the details have been added, assign the **Roles** to the newly created user. The assigned roles will decide the level of access the **User** has in the system.

In this subsection, we have learned the step-by-step process to add a user to the SAC landscape. SAC also allows adding multiple users in bulk by importing a list of users. Let us learn the step-by-step process for importing users in the next subsection.

2. **Creating users by import option:**

SAC also allows Importing multiple users at the same time using the same information we entered while creating a user individually. Having a csv file with the information, it can be imported into SAC and the user information will be stored in the platform.

The entire process of importing users into SAC is shown in Figure 9-9. This option is especially useful when multiple users have to be set up in SAC.

Figure 9-9. *Step-by-Step process to create users using Import option*

Step 1: From the **Main Menu**, click on **"Security"** and then on **"Users"** as shown in Figure 9-9.

Step 2: Click on **"Import Users from File."** Select a File from with the list of users. This file holds the user information in the same format as the User Creation window. This is shown in Figure 9-9.

Step 3: Click on **"Create Mapping"** as shown in Figure 9-9.

Step 4: Map the information in the file with the appropriate parameter of the user creation window.

Step 5: Click **"Ok"** as shown in Figure 9-9.

Step 6: Check if the status shows **"Mapped"** as shown in Figure 9-9.

Step 7: Click **"Import"** as shown in Figure 9-9 above to import multiple users into the system.

In this subsection we have learned how to create multiple users in bulk by importing a list into the SAC landscape. In the next subsection, let us learn the step-by-step process to create Teams for the users imported.

3. **Creating teams:**

Teams allow aggregation of Users into logical groups. For example, a Team of users who is authorized to administer the system would be grouped under the Administrators Team. Likewise, a group of users who frequently access Sales Reports would be grouped under the team of Sales.

Let us now learn the step-by-step process of adding Users to a Team.

Figure 9-10. *Step-by-Step Process to create Teams in SAC*

Step 1: From the **Main Menu** click on **"Security"** and then on **"Teams"** as shown in Figure 9-10.

Step 2: The **Teams** window comes up as shown in Figure 9-10. This window shows the Teams already available within the SAC landscape. Click on the "+" symbol to create a new Team.

Step 3: In the **"Create Team"** window, create a Team by depicting a name, folder, and adding **Users** to the **Team**. As shown in Figure 9-10, create a **Team** called **"Administrators"** with access to the folder **Administrators**.

Step 4: Click **"OK"** as shown in Figure 9-10.

Step 5: The new Team is now visible in the **Teams** window.

In this subsection, we have learned the step-by-step process to create Teams from the Users. In the next subsection, let us learn the step-by-step process to create Roles.

4. **Creating roles:**

Roles define the group of permissions a user has in the SAC landscape. A user once created would need to be assigned the appropriate roles to enable the right level of permissions. Let us now learn the step-by-step process to create roles in SAC.

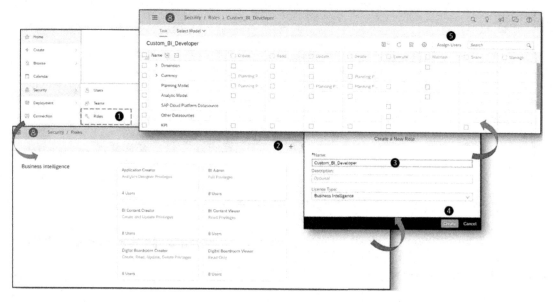

Figure 9-11. *Step-by-Step Process to create Roles*

> Step 1: From the **Main Menu**, click on **"Security"** and then **"Roles"** as shown in Figure 9-11.

> Step 2: The **Roles** window comes up. This shows the **Standard Roles** available for each license. In Figure 9-11, Standard Roles for Business Intelligence are shown. Click on the "+" icon to create a new **Role**.

> Step 3: Fill in the details for the new **Role** such as, "**Name,**" **"Description,"** and **"License Type."** The **License Type** decides whether the user would have access to which type of License, for example, **Business Intelligence** or **Planning**.

> Step 4: Click on **"Create."** The new **Role** is created.

Step 5: For the new Role, Permissions would need to be assigned. Select the Permissions to be assigned to the Role from the available list Role. This can be done by ticking on the "✓" as shown in Figure 9-11. Click on **Assign Users** to add available Users to the newly create Role.

Step 6: Click on ▦▾ to Save the permissions assigned to the Role as shown in Figure 9-11.

We have seen the step-by-step process to create a new Role and add Users to the new Role manually.

Rights can also be assigned by using a template from already existing Roles. If there is already an existing Role, it can be used as a template for the new Role. The template can be modified as per the changes in permissions for creating the new Role.

Let us now learn the step-by-step process to create a new Role using a template.

Figure 9-12. *Create Roles from Template*

Step 1: From the **Main Menu**, click on **"Security"** and then **"Roles"** as shown in Figure 9-12.

Step 2: The **Roles** window comes up. This shows the **Standard Roles** available for each license. In Figure 9-12, Standard Roles for Business Intelligence are shown. Click on the "+" icon to create a new **Role**.

Step 3: Fill in the details for the new **Role** such as, "**Name,**" **"Description,"** and **"License Type."** The **License Type** decides whether the user would have access to which type of License, for example, **Business Intelligence** or **Planning**.

Step 4: Click on **"Create."** The new **Role** is created.

Step 5: For the new Role, Click on ▦ to select from the Template as shown in Figure 9-12. Once the Role is selected from the template, all the Permissions are assigned to the new role. Additional Permissions can be assigned by ticking on the "✓" as shown in the figure. Click on ▨ to Save the Permissions assigned to the Role as shown in Figure 9-12.

SAC also allows Users to request additional rights for the user ID. The Administrator has the rights to approve or reject the requests.

In this subsection we have learned how to create a role from a Template. In the next subsection, let us learn how SAC adds an additional layer of data security by enabling data monitoring.

Data Monitoring is the process of periodically evaluating the data within a data processing landscape to ensure that the data meets the quality and regulatory standards. Data monitoring in SAC ensures that the data loaded into models is as per the standards defined by the organization. Data monitoring is crucial to ensure incorrect data is not sent to the models or stories.

5. **Monitoring Data Changes:**

SAC allows tracking of changes to data in models. This helps track any unexpected data changes and also to check if the data has been changed appropriately.

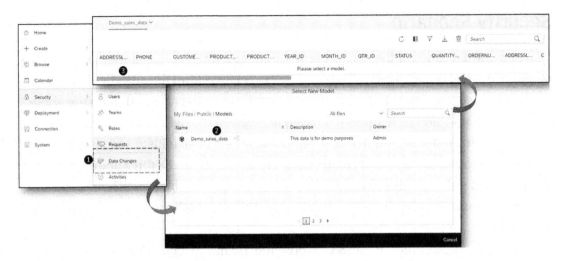

Figure 9-13. *Step-by-Step process to track data change*

Step 1: From the Main Menu, Click on **"Security"** and then on **"Data Changes"** as shown in Figure 9-13. This brings up the monitor screen with the option to select a model.

Step 2: Select the model for which data changes need to be monitored. We have selected the **"Demo_Sales_Data Model"**.

Step 3: All the fields of the model come up. Any change made to the data of the model **"Demo_Sales_Data"** is now liable to be tracked and monitored as shown in Figure 9-13. Any changes to the data are tracked and storied in the above list, and any untoward changes to the data can be easily pinpointed and explored further.

We have now seen the step-by-step process for creating a secure landscape for SAC. In the next subsection, let us learn to implement the concepts and the step-by-step process we learned in the previous sections to create a sample security scenario.

Security Case Study

In the previous sections, we have learned the step-by-step process to create Users, Teams, and Roles. Let us now learn to implement the knowledge assimilated in the previous subsections with an illustration on Role definition for a BI Developer and a Sales Manager in the Retail Business of ABC Inc.

Security Scenario

As we have appreciated at the beginning of the chapter ABC Inc. has multiple businesses ranging from real estate to retail. The primary expectation from the analytics landscape at ABC Inc. is to have data restriction throughout the business roles. Additionally, there are administrators and developers who need to be given the right level of access.

The Administrator or Admin role should have access to all folders and also the permissions to assign rights and create users and teams. All of the three above roles should have a unique definition and be capable of fitting into the overall landscape. Each role should be independent of the other and should fulfill the tasks assigned respectively.

Tom is an administrator in the SAC landscape. He has to regularly monitor the landscape and manage users and roles. As a weekly activity, he also has to make backups of the entire landscape and ensure that the landscape is working optimally without any unexpected errors in the existing stories.

A developer should have access to all datasets and models. The developer should also have the access to change the model as well as the story with unrestricted access across all business functions. Since SAC is devised for being the tool of choice for self-service analytics, the Developer role could be from the IT team or the Business Analyst team comfortable with building stories or changes to the models.

Dick is the developer from the IT department who is actively involved in developing or modifying existing stories and models. As per new requirements from the business, he is also assigned the development of new models and stories for existing and new businesses. In his current engagement, he has been assigned modifying one story from the Sales line of business as a priority. He has subsequently also been assigned the creation of a new story for an existing model in the Manufacturing line of business.

The Sales function is distributed across multiple businesses from retail to real estate. The Sales Manager from the retail business should have access only to sales data from retail and should be restricted from accessing data across other business functions as well as other businesses. The retail business Sales Manager thus should have any level of access across finance or manufacturing and should be restricted to viewing reporting or analyzing data only from the retail sales dataset.

Harry is the Sales Manager from the retail business who would like to access Reports and Stories regularly to be abreast of his performance and to make timely decisions on the next potential customer.

Let us now learn to build a security landscape for the problem defined above.

Solution for the Security Scenario

The security scenario defined above has to be solved as a combination of two levels.

1. Role-level security

2. Folder-level security

We have learned about the process of creating the above security levels in the section "Creating Security in SAC." Let us implement our learning from the section to create the security levels as defined above.

Let us now learn each of the security levels in detail and combine them to form the final security model.

Role-Level Security

In the current security scenario, there are three distinct roles. Let us understand about each of these roles further.

Admin

The Admin or the administrator role has the highest level of access. The Admin should at least have read and write permissions across all objects within the landscape including models, stories, and security objects. Additionally, the Admin will also be able to monitor the landscape, create users and roles, and should also have permissions to delete objects across the landscape. Let us hence assign the standard role to the Admin role by assigning the role **"BI Admin."** Let us then assign Tom to the Admin role.

This is as shown in Figure 9-14.

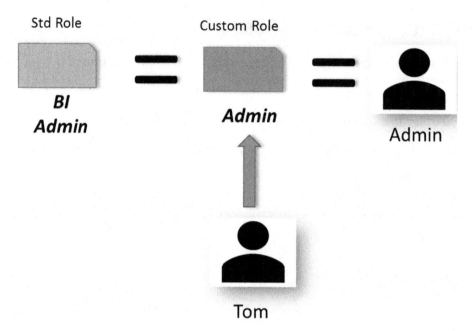

Figure 9-14. *Create Admin Role*

Developer

The next level of access will be with the Developer. The Developer would not have the privileges to change or manage the user security or the system. The Developer would however be able to develop content in terms of Models and Stories. The Developer would also have the permissions to publish the content. If there is a requirement to create a new BI Model or a Story, the Developer should be able to fulfill the requirement. However, the Developer should not have the Permissions to manage users or to monitor the system as in the case of the Admin.

Let us hence assign the standard role "BI Content Creator" to the Developer role and assign Dick to the Developer role.

This is shown in Figure 9-15.

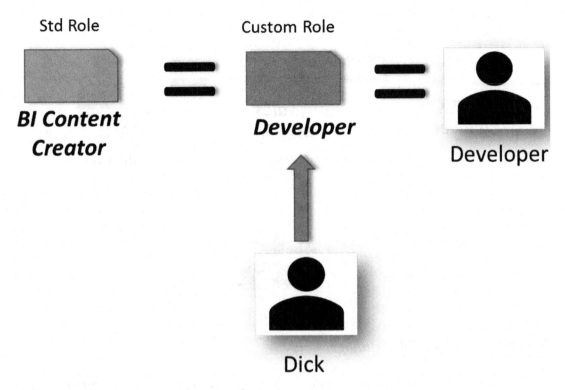

Figure 9-15. *Create Developer Role*

Sales Manager

The Sales Manager would be more of a viewer for the content created by the Developer. Primarily interested in the numbers being put forth by the story created for Sales, the Sales Manager should not have any access to creating or modifying the models or stories created by the Developer or to the Security objects managed by the Admin. In case an incorrect privilege is assigned to the Sales Manager Role, which is typically a business function, there could be a risk of incorrect modification to the model or the story. Hence the business role should not have any other access except viewing and sharing.

Let us hence grant the role "BI Content Viewer" to the Sales Manager Role. Now let us assign Harry to the Sales Manager Role.

This is shown in Figure 9-16.

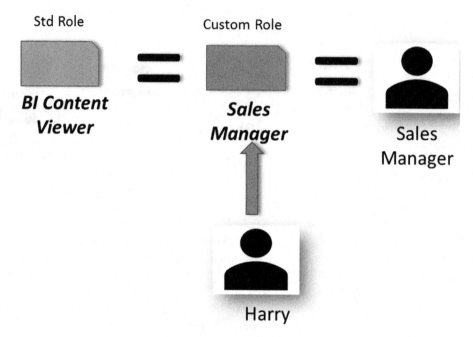

Figure 9-16. *Create Sales Manager Role*

Note that we have used standard roles for creating meaningful roles within the SAC landscape. For example, though the standard role is BI Content Viewer, the same role can be assigned to Sales Managers or Finance Managers or any other business function. Further security can be assigned on the folder level, allowing end users of business functions to only access content relevant to their realm of business. Having custom roles instead of standard roles enables ease of use across the landscape while also enabling the Admin to ensure the landscape is clean and not cluttered with multiple users being assigned to the same role. It is easy to decipher that the "Sales Manager" role stands for the sales function, instead of "BI Content Viewer." The Sales Manager role can easily be assigned to only the relevant content by assigning folder-level security.

Let us learn about folder-level security in the next subsection.

Folder-Level Security

In the previous section "Role-level Security" we learned the process of successfully creating roles and assigning users to the roles. In this section let us learn to assign the folders to each of the security roles created and understand how folder-level security can help users in their specific roles to carry out their tasks diligently.

Admin

Since the Admin is a very powerful role within the landscape, the Admin should have visibility in to all the folders within the landscape. Once the user is added to the Admin role, one of the default rights allows the user to have access to all folders within the landscape. The Public folder will have subfolders that the Admin will have default access to. We have already assigned Tom to the Admin role.

As can be seen from Figure 9-17, Tom will have access to folders for Sales, including a subfolder for Retail, Finance, and Manufacturing.

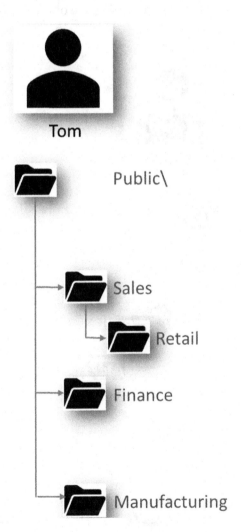

Figure 9-17. *Folder Structure for Admin user*

Developer

The Developer role will have access to only to those folders for which permissions have been granted. Within the folders, the developer will have full access to create and modify stories. If there is a Models folder for saving the models separately, the developer will also have access to this folder. However, this depends on the organizational folder structure.

We have already assigned Dick to the Developer role. Tom will thus have access to folders for which access has been provided, for example, Sales including the subfolder for Retail and Manufacturing. The other folder, Finance, will not even be visible to Dick hence restricting his access to the data as well as the objects within.

This is shown in Figure 9-18.

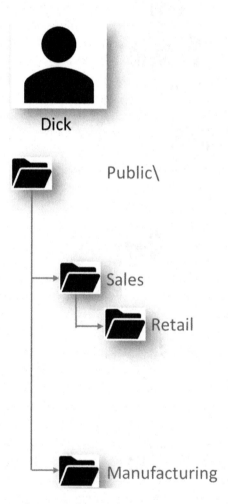

Figure 9-18. *Folder access for Developer User*

Sales Manager for Retail Business

We have already created the Sales Manager role and assigned Harry to the role. However, Harry should not have access to the entire Sales line of business. His access should only be restricted to his area of interest, which is the Retail Business. Creating a separate role for only Harry would not be a prudent approach to maintaining security. A mechanism is hence desired, which should enable Harry to be able to access all the reports and stories for sales but only for retail business.

This can be achieved by creating a subfolder for Retail under the Sales folder. The Retail subfolder will inherit all the permissions assigned to the parent folder while allowing assignment of roles to the subfolder. Let us grant access to Harry to the subfolder Retail. Hence Harry will have all the permissions of the Sales Manager role and then to the subfolder Retail, thus, fulfilling the requirement.

This is shown in Figure 9-19.

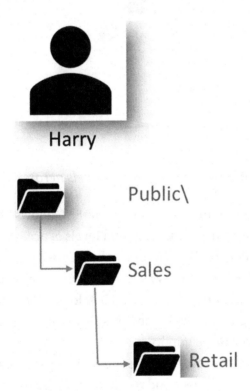

Figure 9-19. *Folder Access for Sales Manager Role*

Once all three users have been granted the security as desired, a comparative access level of all the three users with varying levels of access to the folders as seen in Figure 9-20. This ensures that the users have minimal difficulty in navigating the landscape by allowing the users to see only the folders they have access to, while also maintaining security. The comparative analysis is shown in Figure 9-20.

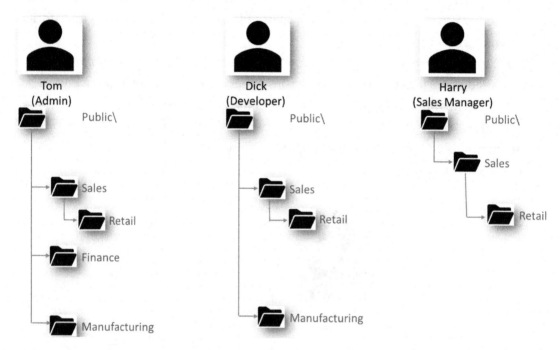

Figure 9-20. *Comparative analysis of access levels across roles*

We have now seen an example scenario on how SAC allows flexibility in security settings by allowing multiple users with different levels of access to be created and maintained. We have understood how roles can be created and users assigned to standard and custom roles. We have also learned about folder structures and how to create subfolders while assigning users to specific folders. SAC allows for multiple combinations to be created for specific use cases for security. Hence creating a robust security model gains all the more importance.

All of the above features ensure that ABC Inc. can fully utilize the native security built within SAC and save significant costs and resources in implementing security across the new analytics landscape.

Summary

And this brings us to the end of this chapter on "Design A Secure Platform Using SAC." In this chapter we have learned how to implement security within the SAC landscape and its importance in the Analytics Road Map for ABC Inc. In this chapter we have learned about how SAC provides a robust architecture across multiple facets of the security. We have also learned about the following:

1. The importance of having a secure Analytics Landscape.

2. Step-by-Step process to design and deploy a security architecture in SAC.

3. Customer benefits of SAC's Security architecture.

4. Future road map for SAC's security architecture.

In the next chapter, let us learn about the top trends in analytics and how SAC is aligned to these trends. Let us also learn how SAP releases regular updates to SAC to enable enterprises to improve rapid decision-making and self-service while continuing to build new capabilities in accordance with the new trends.

CHAPTER 10

Product Road Map and Future Direction for SAC

The pace of innovation and evolution of technology compounded with a dynamic business environment make it absolutely essential for every organization to understand and manage IT and associated risks that are incident upon and their implications for business. With an ever-growing adoption and application of technology across enterprise functions, the need for awareness becomes fundamental to managing IT. One of the best ways to minimize risk and optimally gain from investments in technology is having better awareness and clarity on the technology road map and its direction of evolution. Understanding the road map and future direction help mitigate the risks emanating from a possible gap as well as aligning business plans and strategies with the forthcoming capabilities and changes in the technology. Having an eye on the product road map also ensures that all stakeholders are mindful of upcoming changes and being upskilled too.

Evolution of any Product is an outcome of multiple factors, with the key being its use and adoption, the components that make it. Innovative techniques for data storage have emerged at a rapid scale. The terminology of storage has grown so wide that figures like zettabytes or domegemegrottebyte bytes are no longer imaginary storage figures. As the technology is evolving, SAC will continue to integrate the latest developments for delivering more value from its use and adoption. Use and adoption could translate into interface or augmented technologies or application to existing or new problems; similarly, its component could mean storage or processing technology or the algorithms and models or the architecture.

© Vinayak Gole, Shreekant Shiralkar 2020
V. Gole and S. Shiralkar, *Empower Decision Makers with SAP Analytics Cloud*,
https://doi.org/10.1007/978-1-4842-6097-5_10

Appreciating the critical significance of understanding the Product Road Map and Future direction of technology to any enterprise, the content in this chapter will begin with observing key trends in Analytics technology with brief references to the top trends. We will then learn how SAC's capabilities are aligned with these trends. Further on, we will explore trends that are shaping the Industry and how SAC's capabilities are mapped to them. We will then understand SAP's vision for the Experience Company powered by the Intelligent Enterprise, and how this aspiration of intelligent technologies to power the enterprise shapes and triggers developments in SAC. We will learn about the primary components of SAP's vision and how SAC comes built in with technologies both from experience and intelligence components. Finally, we will learn about the road map for SAC and how to stay updated about it.

Top Trends in Analytics and SAC's Alignment with the Trends

In the past decade, Business Intelligence that primarily worked with structured data has evolved into modern analytics that we know of today, works with a variety of data, for example, unstructured, streaming, and big data along with structured data. For the next decade, further enhancements and paradigm shifts in business analytics will involve all the new age technologies like graph analytics and blockchain. Let us learn about some of the top trends in Analytics in the coming decade. We shall also learn to identify how trends are shaping SAC and its road map.

1. **Augmented Analytics:**

Augmented analytics is the use of machine learning and AI to assist end users with data preparation and exploration. Augmented analytics enables end users to deliver faster insights to action by reducing the time taken to deliver action on specific outcomes of data analysis. Redundant work across multiple layers of hierarchies is time consuming and results in delay of the crucial and timely decision-making process. Augmented analytics reduces this redundant effort and enables rapid decision-making. Redundancy in work is reduced considerably by automating the data preparation and delivery process. Augmented analytics also enables end users to rapidly scale up on advanced analytics with a low learning curve. This enables organizations to take advantage of advanced data analysis techniques while investing less time and effort in the learning process.

With the exponential rise in data volumes, the demand for rapid analysis of the data has also been rising. Additionally, as organizations adopt automation across the enterprise, demand for automating data management and reporting has been one of the key priorities. This has led to the demand for software vendors to build in augmented analytics features across the spectrum of data management including data quality, master data management, and data integration practices. Conversational AI experiences coupled with algorithms for rapid data processing enable end users to scale up data analysis tasks to deliver real-time business decisions. Augmented analytics including automated data discovery, data management, and conversational AI using NLP and visualizations is slated to be a prominent component for analytics applications and landscapes.

Intelligent technologies like machine learning and augmented analytics form the core of the future vision that SAP has for SAC. Using a combination of natural language processing and newer visualizations through the new Smart Insights can enable end users to mine data and find hidden trends without having exposure to complex data science techniques or tools. Smart Discovery delivers ready-to-use stories with built-in analysis of influencing factors and what-if analysis by changing the influencing factors.

Smart Predict will see the transition from black box AI to explainable AI for increased explainability for time series models. Support for live connectivity will enable the models to work on live data not stored on the SAC landscape.

2. **Blockchain in data and analytics:**

Blockchain has been finding use cases across the entire spectrum of the IT landscape. Transactional systems have been reliant on the distributed ledger for delivering trust across transactions. Data and analytics technologies with integrated blockchain will enable in keeping data lineage and transparency across multiple consumers of the data.

3. **Commercial AI and Machine Learning:**

Currently AI and ML algorithms are dominated by open source platforms with little or no contribution from the commercial players. For an enterprise to remain competitive, it is of utmost necessity to have AI and ML algorithms integrated into commercial platforms, both transactional and analytical. Having commercial software platforms provides AI and ML functions enabling client enterprises to have better support and a clear vision for the road map of development for the product. This will also ensure a singular experience in terms of reuse, model management, and deployment of enterprise-scale algorithms.

Artificial Intelligence is real and has already found its way into multiple applications. Built as black box ready-to-use algorithms and modules, AI enables applications to deliver rapid outcomes and be agile and automated, enabling organizations to have lower time to value realization cycles. As AI becomes commercialized and rapidly customizable, embedded AI would enable analytics to be more reliable and rapid. End users would be able to rapidly narrow down to the data points of interest augmented by Intelligent technologies working in tandem in the background. Explainable AI would also simplify the complicated AI model and enable end users to be able to justify their findings with clear and to-the-point explanations.

SAP's vision for the intelligent enterprise is based on the pillars of intelligent technologies, IOT, and Analytics. Available across both the transactional and analytical landscapes, intelligent technologies deliver a uniform enterprise scale AI and ML experience to end users. And SAP has been investing on building these technologies further by delivering Smart Predict enhancements in SAC by refining the already available scenarios and delivering further scenarios for predictive technologies. Additionally, Smart Discovery, which enables auto creation of stories, will see better support for wider KPIs.

4. **Continuous intelligence:**

The emergence of tools to enable real-time data availability along with streaming analytics have enabled organizations to benefit from continuous intelligence. Enterprise customers can work with data both structured and non-structured delivered rapidly to the analytical landscape to deliver trends and insights powered by AI. Continuous intelligence is of utmost value to businesses for rapid and structured decision-making.

SAP's road map for SAC brings together an enhanced modeling experience built on Agile concepts.

5. **Data Fabric:**

Datasets are built not just from standard data sources but also as a combination of data pipelines, semantic layers, and APIs. A data fabric enables rapid and efficient movement of data across the analytics landscape, enabling end users to pick and choose datasets and come up with the best results.

Contextual Analytics enables seeing KPIs and metrics in a completely new light. However, enabling the context for the numbers in dashboards and stories requires multiple data sources to be integrated and processed. As multiple sources emerge, integration becomes a complicated process. Data Fabric enables a combination of data pipelines, APIs, and other traditional data sources combined to deliver a wholesome, reusable source for the reports, stories, and dashboards. Data Fabric enables analytics tools to be more agile and wholesome in having a single data source with reusable components.

6. **Explainable AI:**

Most of the commercial AI solutions are available as black box products. Though these solutions fare very well in delivering results, without the ability to explain the outcomes, the trustworthiness of the algorithms becomes a risk. Explainable AI drives transparency and trustworthiness of the AI solutions enabling outcomes to be explained and explored. The entire process can be easily comprehended instead of just relying on the results.

SAP's road map for Smart Predict will see the transition from back box AI to explainable AI for increased explainability for time series models. This will enable end users to be reliant on the AI and deliver predictive outcomes that can be fully explained and comprehended.

The time series forecast currently can be segmented only across one dimension. With the future releases, the time series forecast will be enabled for multiple dimensions. This will enable multidimension forecast analysis enabling businesses to implement decisions contextually across variables and signals.

Usually with machine learning technologies, certain results have to be arrived at with numerical analysis with no clear explanation as to the results. Models trained over time derive enough confidence to learn and interpret data and signals and deliver presumably correct results.

Explainable Artificial Intelligence on the contrary enables human experts to understand and interpret not just the results but also the underlying logic behind the same. With these technologies, business decisions involving Artificial Intelligence enable decisions to be more confident and build trust.

SAP recognizes this trend in Artificial Intelligence and machine learning technologies and as part of its road map for SAC has plans to integrate the same. This will enable organizations to embrace AI technologies and be confident of the outcomes and the business decisions it supports. This will enable less iterations over training of the AI model and can be sent for production at the earliest with clear understanding of the way the model has delivered results.

7. **Graph Analytics:**

Graph analytics enables discovery of relations between entities, for example, people and places. Data has been growing exponentially and the edge gained by data analysis has been no less of a necessity in the competitive business world. End users demand answers to complex business queries built across structured and unstructured data including blending and wrangling data across multiple applications, which in most cases is beyond the capabilities of the traditional SQL and data warehouse model.

And it is here that graph analytics finds the most use cases including route optimization for logistics and fraud detection. As per Gartner, a leading research firm, the graph databases are slated to grown at 100% over the years.

8. **Persistent memory servers:**

Current database technologies have incorporated in-memory database structures, delivering tremendous value for real-time decision-making. However, with increased data loads, memory might not always be available for rapid data processing. Persistent memory technology enables rapid data processing delivering actionable insights. Though still in the nascent stages, persistent memory servers for database management systems will enable the next stage of advanced analytics.

SAC is built natively on the HANA cloud platform, which has been one of the first database systems to actively utilize persistent memory technology. In tune with the technology advances, on the database side, SAC will have better data processing capabilities along with increased connectivity to both on-premise and cloud sources.

In this section we have learned some of the top trends for Analytics and how SAC's capabilities road map is aligned to the most relevant ones, that is, augmented analytics, explainable AI, continuous intelligence, and persistent memory. We have understood the most important factors that will be shaping how analytics as a technology is perceived and how these very factors would shape the future implementations of analytics.

In the practical scenario, industries might not yet be ready to accommodate these trends into their landscapes. Some of the trends might still be in the developing phase and might not be fully capable of delivering business value to organizations or the entire industry as a whole. Additionally, there could be other factors and technologies that will enable the industry to bring additional value into the way analytics projects are implemented to bring in value. For example, analytics trends might suggest blockchain for data and analytics; however, the industry might not yet have reliable products supporting the technology. On the contrary, storage technologies that have evolved rapidly might add true business value by enabling storage of large quantities of data for further processing and analysis.

In the next section, let us appreciate recent industry trends in technology and how SAC has been adapting to these trends to come up with features aligned to technology trends.

Industry Trends and Alignment with SAC Capabilities

Trends in the Industry influence its evolution, which in turn shape the trends in analytics and data. With a view to relate the evolution of SAC to the trends in analytics, in the following section, we will learn about trends in the following areas:

- Use and adoption

- Components and technology

Let us also explore their influence on the product road map of SAC.

1. **Use and adoption:**

Analytics can deliver maximum value for business and enable end users to deliver strong business decisions by making curated data available for self-service and analysis. Business users need to be empowered with analytics and should be encouraged by adopting self-service capabilities in addition to standard reports available for information about the available data. Let us learn the components in the use and adoption capabilities.

1. **Self-Service:**

Technology is seeped into society so early and deep that each professional is adept at learning and keeping pace with basic technology, and this has manifested into self-service expectations and specifically into Analytics so that a line-of-business professional can perform queries and generate reports with no or nominal IT support. The screen, screen elements like menu path, and tabs as well as schema are crafted for easy adaptation by an average human being.

Self-Service Analytics is of prime importance for business analysts as well as end users who want to analyze data seamlessly to discover trends and arrive at business decisions rapidly. Enabling end users to handle data analysis reduces dependency on IT teams for organizations and promotes agility and accountability.

2. **Augmented Analytics:**

Augmented analytics is the use of machine learning and AI to assist end users with data preparation and exploration. Augmented analytics is one of the top 10 analytics trends identified by Gartner.[1] Augmenting self-service with assisted AI and machine-learning capabilities enables analytics and data analysis across all levels of the enterprise hierarchy. Reducing the learning curve steeply as also the dependency on IT and data scientists, end users can implement complex data analysis, create dashboards, and analyze trends. Backing business decisions with strong data analysis enables business leaders to take calculated risks while gaining the edge over competition.

2. **Components and technology**

Technology has rapidly evolved to be aligned with human evolution itself. In accordance with Moore's Law, computing power has steadily risen while costs have equally steadily dropped. Cloud technologies have brought about a new paradigm by enabling unlimited flexibility and scalability over shared resources in data centers. Advances in storage technologies have seen the emergence of in-memory and solid-state devices allowing rapid data storage and retrieval. The advances in technology have enabled organizations to delve deeper in to both structured and unstructured data, transform it, and explore it for trends and cycles.

Let us now understand some of the prime factors that have fueled the technology evolution to progress rapidly in the recent years.

---

[1]https://www.gartner.com/smarterwithgartner/gartner-top-10-data-analytics-trends/

1. **Storage Technology**

   The emergence of magnetic disks to store data marked the first major step in the evolution of storage technologies. Flash technology enabled storing of data into memory cards or drives. Though flash drives enabled rapid storage and retrieval of data, capacity has always remained restricted. Recent innovations have enabled arrays of flash drives to store data and enable cloud-capable data storage.

   a. **Cloud:**

      Cloud technologies have in recent years emerged as the cornerstone for enterprise technology landscapes. Offering rapid scaling and flexibility with minimal investments, cloud technologies offer storage and compute solutions that are ideal for processing large volumes of data. Working on a distributed architecture with algorithms for consuming analytics, cloud technologies have emerged as the instrument of choice for heavy-duty analytics.

      Cloud-native solutions are built over the cloud and can be accessed over the internet. SAC is one such application from SAP that has been built ground up on the SAP Cloud Platform. Cloud-native applications are built for high flexibility with a modular design, which makes them easier to integrate and add new features. They are also better attuned to deliver high performance owing to the modular nature and being built to best utilize the features of the available cloud infrastructure. The development of SAC has been in line with SAP's vision of having focused on the cloud platform for delivering continuous integration and innovations.

   b. **In-memory Computing**

      A standard analytic report just like any other application executes by pulling values from the storage, executing in memory, and sending back the data to the storage device.

The number of fetch and store actions add to computing time resulting in significant delay in the overall execution of the application. In-memory computing also allows massively parallel operations enabling faster outputs to data operations.

In-memory computing reduces the number of fetch and store actions enabling applications to deliver high performance. SAC is built on the SAP Cloud platform and works over HANA as a database management system, which is a robust in-memory database. SAC is able to deliver high-speed analytics using the underlying HANA database. Further enhancements in SAC would be delivered to effectively utilize the power of HANA more effectively to deliver the analytics.

2.   **Streaming Technology**

With the availability of scalable hardware and computing power, the ability to analyze and derive meaning from a continuous stream of data has emerged as a popular analytics ability. Streaming analytics enable real-time decisions especially for analyzing IOT data wherein data can be analyzed rapidly for any anomalies in the machines.

3.   **Edge Computing:**

Edge computing is a distributed system that does the data processing near the source of the data. Termed as the next generation of computing post the IOT and cloud era, edge computing strives to provide true real-time analysis of data rather than sending the data to a processing unit in a server farm or on a cloud data center. Edge computing will drive the internet of everything with things and people both creating and consuming the available information. The name is derived from the fact that the data is analyzed at the Edge of the network before being sent to further processing or storage on the cloud or on-premise data centers.

4. **Processing Technology:**

Computing power of processors has multiplied rapidly in accordance with Moore's Law. Data processing capabilities have evolved in alignment with computing power. However, along with the hardware, recent developments in technology have seen the rise of massive parallel processing and distributed processing of data. Google's "MapReduce" algorithm for processing large quantities of data and the evolution of Hadoop for Big Data processing have enabled enterprises to utilize the full capacity of available data.

5. **Statistical Models:**

Statistical models have been prevalent across technology domains especially for advanced forecasting and mathematical analysis. The availability of powerful hardware combined with best-in-class processing capabilities have seen statistical models being available across all roles within the organization. Black box AI enables end users to build predictive models and analytics without being exposed to underlying complexities. Predictive Analytics and AI are core capabilities of modern analytics landscape and enable informed decision-making capabilities.
SAC's Smart Predict delivers pre-built scenarios for classification and regression, and time series algorithms deliver the easy and ready-to-use predictive capabilities to end users. SAP has been constantly upgrading the Smart Predict feature for bringing in better explainability to the algorithms already available with plans to include further models into the Smart Predict portfolio.

6. **Business Content and Industry-Specific Analytics**

End-to-end analytics covering scenarios across all aspects of a particular industry provide customers with a starting point in their analytics journey. By rapidly deploying industry standard solutions and connectivity to standard transactional sources, business content enables customers to scale up rapidly to

advanced enterprise analytics. Industry-specific analytics
deliver predefined KPIs and metrics for enterprises to track
within an industry and deliver the most-relevant dashboards
to the business.

SAC provides ready-to-deploy business content for over 40
LOBs and industries with new content being released steadily
every quarter. The complete list of the business content is
available at `https://www.sapanalytics.cloud/learning/`
`business-content/` that is updated regularly and can be
checked for new content released.

In this section we have learned the influencing factors for industry trends and how
SAC's capabilities are mapped to these trends.

In the next section, let us learn about how analytics forms a core component of
SAP's vision of the Intelligent Enterprise. Let us also learn how ABC Inc. can take the
first step to moving to the Intelligent Enterprise by employing the capabilities of SAC for
delivering maximum business value.

We had appreciated the challenges and difficulties faced by ABC Inc. in Chapter 2
that were later resolved by specific capabilities of SAC, in Chapters 3 to 9 and learned the
step-by-step process to achieve the requirements of ABC Inc. within the SAC landscape.
In the following section, let us explore how the evolution of SAP and primarily SAC can
enable ABC Inc. to develop into a modern Intelligent Enterprise and deliver cutting-edge
analytics to end users. We will also learn how this will be the foundation stone for setting
up the Intelligent Enterprise at ABC Inc.

SAP's vision for the Intelligent Enterprise

SAP has been the market leader in Enterprise Applications with over 440,000 customers
across 180 countries. SAP's ERPs have led the world of business applications for nearly
four decades. Over the years, SAP has consolidated its position not only in the business
applications space but also in other technologies. With SAP Business Warehouse, SAP
has delivered a robust Data Warehousing Application. With the acquisition of Business
Objects, SAP consolidated the leadership position in Business Reporting technology space.
The SAP HANA database has now matured into a full-fledged data management system.

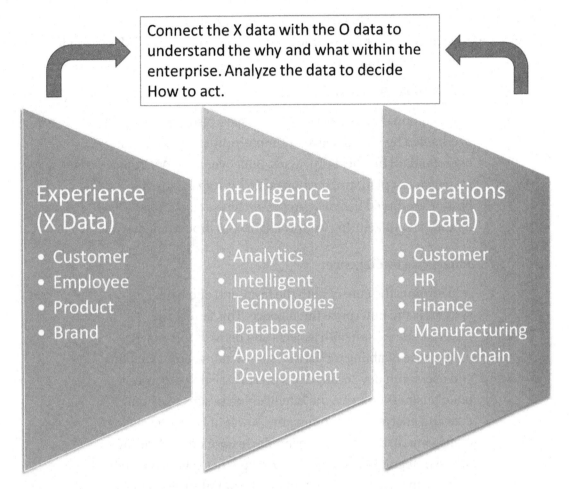

Figure 10-1. *SAP's vision for The Experience Company*

1. **The Experience Company driven by the Intelligent Enterprise:**

User Experience has been at the forefront of SAP's strategy, which has seen the emergence of a powerful new UI. SAP's acquisitions in recent times as well as its strategic solutions have all been enabled with the powerful UI5 technology, which focuses purely on end-user experience. With vibrant colors and responsiveness across devices, UI5-based technologies have been delivering a best-of-the-class experience across customer, employee, product and brand. All of these front-end solutions have a powerful one-on-one relationship with the end user.

SAP Fiori allows for vibrant front ends to be crafted along with the tile-based screen that forms the entry point to any of SAP's products. The experience across all of these products is uniform and consistent, ensuring the end user has the same experience across all of SAP's products.

SAP's recent acquisition of Qualtrics enables organizations to further tweak the customer experience and ensure the brand equity always stays high. SAP's vision for the Experience Company is illustrated in Figure 10-1.

1. **Operational Systems:**

   The traditional transactional systems have also been overhauled by SAP to deliver a simple, robust model working over standard business processes. Built over the HANA data management platform, the transactional systems provide best-of-the-breed business processes across customers, HR, finance, manufacturing, supply chain, and procurement.

2. **Enterprise Intelligence:**

   And forming the core of both the front-end and back-end systems is the Enterprise Intelligence. This landscape integrates data from both aspects and delivers top-notch intelligence enabling robust decision-making. From enabling end users with day-to-day decisions to assisting executives build business strategy, the Intelligence systems work on data and analytics. Combining the power of HANA for data management, these sets of technologies work on delivering a robust, integrated, intelligent landscape that acts as the brain of the both the extremes of the Intelligent Enterprise. One of the components of this landscape is Analytics.

   Driven primarily by SAP's vision of a single platform for all analytics, the analytics component comprises of Augmented technologies, collaborative planning, predictive analytics, and modern data warehousing. SAC is at the forefront of analytics by offering a single platform for all analytics over the cloud. Enabling multi-device access and offering a rich UI, SAC also enables augmented analytics, predictive scenarios, and integrated planning capabilities.

2. **SAC's alignment to the Experience-driven Intelligent Enterprise:**

To achieve the vision of an Intelligent Enterprise, SAP has been investing steadily in Intelligent Technologies across the SAP Product Suite. To enhance the experience, SAP has been moving to the intuitive UI5 framework and Fiori. SAC has similarly seen further integration with Smart Capabilities while providing the end user with an exceptional experience. As part of the enhancements for SAC in terms of Intelligent Technologies and User Experience, here are the following components:

1. **Intelligent Technologies:**

   Intelligent technologies like machine learning and augmented analytics form the core of the future vision that SAP has for SAC. Imagine a Smarter Smart Insights backed by a powerful machine learning algorithm. End users can interact with the Smart Insights to develop points of interest from not just charts and graphs but also from tables and grid components as well. A new insight enables tracking changes in data over time and aided by a machine learning algorithm brings up valuable insights as per the changed data. Using a combination of natural language processing and newer visualizations, the new Smart Insights can enable end users to mine data and find hidden trends without having exposure to complex data science techniques or tools.

   Further automation is on the cards with planning operations being powered by automated end-of-planning cycle operations. Structured allocations would enable multiple planning operations to be completed at one go. Data Actions are also being modified to enable asynchronous execution. This will reduce the waiting time for planners and simultaneous operations can be handled using the same data actions.

2. **User Experience:**

   One of the core tenets of SAP's Experience company is the user experience. Newer visualizations with improved chart-building experience will enable end users to rapidly spot trends and drive informed decision-making processes within the organization.

Qualtrics is an experience management company that enables organizations to systematically measure and improve the end-user experience. With various options for customer experience, employee experience, product experience, and brand experience, Qualtrics data enables organizations to accurately measure the end-user experience across all dimensions of an organization. By enabling a Qualtrics connector to SAC, organizations will be able to integrate both experience and transactional data into a central intelligence landscape, which will house 360-degree actionable intelligence.

3. **Anytime Analytics**

As we explored in Chapter 6, SAC enables anytime analytics thorough the native-cloud interface as well as the mobile app. The mobile app enables access to SAP Analytics cloud though with multiple limitations. Imagine an executive who looks at a story on the mobile and feels something is out of place. The future road map would see the mobile map supporting annotations. The executive would be able to immediately highlight the changes in the mobile app and send it back to the appropriate analyst to relook at the trend.

Mobile hardware has rapidly been evolving to enable a powerful handheld device that can power multiple business applications and functions. Analytics and data processing are rapidly shifting to mobile devices to enable end users to explore data on the go without being confined to a desktop device. Mobile Intelligence is currently in a nascent stage by being restricted to the responsive canvas and the mobile app. The future would see apps being at the forefront of delivering Enterprise Intelligence across the entire organizational structure. Real-time annotations, embedded collaboration, and augmented analytics would enable organizations to be agile, quick, and collaborative.

Search to Insight enables interacting with the SAC landscape through Natural Language for sifting through data and finding trends. The Search to Insight augmented analytics application would soon be available on the mobile app, enabling natural language communication with the SAC landscape through mobile. This will be the true embodiment of anytime analytics with end users being able to interact with the landscape from anywhere at any point in time.

4. **All Round Integration:**

As per the SAC road map, live connectivity to SAP data sources will enable enhanced features, including support for custom hierarchies and query changes. Augmented analytics will find more support for live connections including Smart Discovery for building stories.

Import connectivity will see higher thresholds in terms of ingested rows for files.

In this section, we have seen SAP's vision of the enterprises to move the needle to the next level of the experience-driven Intelligent Enterprise. Previously we have appreciated the analytics trends, and then industry trends before learning about SAP's vision for the intelligent enterprise and analytics. In the next section, let us learn the future roadmap for SAC and how to access the road map to benefit the most from upcoming releases with innovations across the trends we have already learned about in sections "Top Trends in Analytics and SAC's Alignment with the Trends" and "Industry Trends and Alignment with SAC Capabilities" in this chapter.

3. **Future Road Map for SAC:**

SAC has been the focus of innovation across Business Intelligence and Analytics for SAP and with new features constantly being released in alignment with major analytics and industry trends. SAP has been constantly upgrading the features of SAC with connectivity to all major data sources and native connectivity to SAP products, both cloud and on premise. The data management and story delivery capabilities are also under constant upgradation. SAC continues to deliver the best-in-class analytics with the ability to also work with the SAP Business Objects Portfolio in Hybrid mode. In this section we will learn about the future road map for SAC and how it is aligned to the overall analytics strategy of SAP. However, before we learn about SAC's road map, let us first learn about SAP's road map for the Analytics portfolio. This will help us understand the context to development of SAC and how it fits into SAP's product portfolio for analytics.

1. **Road Map for SAP Analytics Portfolio:**

    There has been a wider acceptance of analytics as a formidable force in the extremely competitive business world in the last decade. To keep up with SAP over the last decade and acquired and developed multiple technologies, applications and tools to cater to different segments of the analytics consumer community have been developed. However, this resulted in multiple tools across the portfolio with capabilities overlapping across tools and products, prompting SAP to start a rationalization campaign to consolidate the tools across the SAP Analytics Portfolio. The result of the consolidation is shown in Figure 10-2 below.

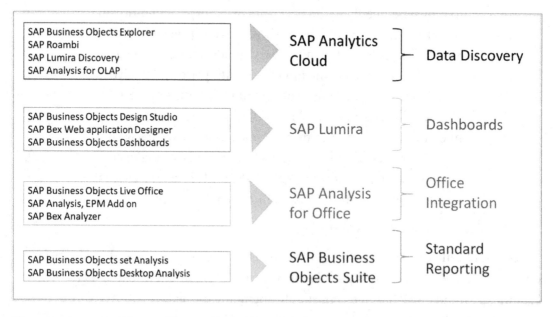

Figure 10-2. *SAP's Product Consolidation for Analytics*

As per SAP's statement of direction for SAP Analytics portfolio, SAP has been focusing on two primary aspects of delivering the best analytics experience to the customer. The entire statement of direction for SAP Analytics Portfolio can be referred to at `https://www.sapanalytics.cloud/wp-content/uploads/2019/11/SAP-Analytics-Business-Intelligence-Statement-of-Direction-Nov2019.pdf`

In conjunction with SAP's strategy, the three approaches are the following:

1.   Standard Reporting

2.   Advanced Analytics

3.   Hybrid Analytics

Let us explore each of these perspectives to understand SAP's vision for SAP Analytics implementation.

1.   **Standard Reporting:**

SAP Business Objects Suite has been the product of choice for enterprise reporting spanning nearly two decades of innovation in delivering reporting processes across the enterprise. SAP's strategy for customers with enterprise reporting needs has been to align with SAP Business Objects. SAP Business Objects provide a robust web-based platform for creating and publishing reports across the enterprise. The SAP Business Objects Portfolio with Web Intelligence and SAP Lumira continues to deliver Enterprise Reporting along with reports and dashboards, including scheduling and publishing.

2.   **Advanced Analytics:**

SAC provides a modern cloud-based platform for enterprises to scale up to new age analytics trends like augmented and predictive analytics. Due to the cloud-native nature of SAC, and subscription-based licensing, enterprises can rapidly scale and be flexible while creating a robust analytics framework for the entire organization. We have discussed the capabilities of SAC to address typical enterprise requirements in Chapters 3 to 9 of this book. SAC enables end users with advanced analytics and applications across the enterprise for delivering self-service analytics and real-time decision-making processes. SAP's strategy proposes SAC as the solution of choice for advanced analytics to the end users.

3. **Hybrid Analytics:**

For customers who are already on the robust SAP Business Objects platform, SAP recommends introducing SAC for advanced analytics. The Business Objects platform enables standard reporting process across the enterprise whereas SAC can be used for real-time analytics, embedded analytics, and in-depth data analysis powered by augmented analytics. SAP also provides a license migration path for moving the Lumira Discovery stories to SAC.

Customers with the hybrid analytics approach have multiple benefits since the current landscape is not disrupted and the new analytics platform can be merged without any major issues. End users can rapidly scale up and start building stories and predictive analytics to supplement the standard report delivered by SAP Business Objects. The underlying integration with Microsoft Office enables end users to easily work with familiar tools like Excel and PowerPoint while building data analytics on the cloud.

In this subsection, we have understood the overall road map for the entire SAP Analytics Portfolio and learned of SAP's vision for standard reporting with SAP Business Objects Suite and advanced analytics with SAC.

In the next subsection, let us learn the specific road map for SAC and SAP's vision to enable advanced analytics across the enterprise.

2. **Road Map for SAC:**

SAP has a quarterly release schedule for SAC with new features being added at regular intervals. To enable customers to plan ahead in terms of new features and plan their development activities in alignment with the plan, SAC provides a detailed road map. The road maps for SAC features are available at `https://roadmaps.sap.com/welcome`. Before we learn about the future releases within the SAC roadmap, let us first comprehend the step-by-step process to access the roadmap.

Note   The road maps and forward-looking statements cannot be considered as commitments by SAP to deliver. Innovations delivered are subject to change.

---

1. **Accessing road maps:**

In this subsection we will learn the step-by-step process to access the SAC road map.

Figure 10-3. *Step-by-Step process to access SAC Road Map*

Step 1: Start the browser and go to `https://roadmaps.sap.com/welcome`. Log in using the S-user id, which is the id provided by SAP for access to the SAP systems. This is as shown in Figure 10-3.

Step 2: The webpage displays the search bar including other information about how to access the road maps. In the search bar, type SAP Analytics Cloud as shown in Figure 10-3.

Step 3: The road map for SAC comes up, which shows the current enhancements and future enhancements expected. The road map is divided into sections based on the quarterly releases as well as future releases. Each release is further grouped in to sections based on the capabilities of SAC.

Step 4: Click on each section within the release to bring up the detailed road map. For example, clicking on Analytics Connectivity in the release of Q2 2020 brings up the detailed features expected in this release. This is shown in Figure 10-3.

In the next subsection, let us understand a summary of the major enhancements for SAC. Note that these are planned enhancements and might or might not be delivered.

2.   **Enhancements road map for real-time decisions:**

Real-time decisions enable end users to reduce the time taken to deliver crucial business decisions based on data. SAP's road map for creating connections and enabling rapid access to data will enable end users to connect to data, build stories, and rapidly make business decisions. Real-time decisions are essential in offering a competitive edge in today's business scenario while also allowing savings in time for users to focus on other important tasks at hand. In this subsection, let us learn the enhancements planned for SAC for enhancing the capability of real-time decision-making.

Connections

Connections form the core for bringing in data to the SAC platform. Live connectivity enables SAC to connect to the back-ends applications and deliver data for end users to analyze for rapid decision-making. Backed by data and predictive technologies, structured decisions are based on data and not past experiences.

The future direction is to enable other SAP sources to connect to SAC over live connectivity. This would further help in reducing the data footprint.

Import connections offer flexibility and deeper data analysis. SAP's future plans include connectivity to a wider range of data sources, including non-SAP data sources. SQL-based databases would also be available as import data sources.

With multiple improvements in connectivity technology, SAP plans to upgrade SAC to be more resilient in connecting with the back-end applications. Hierarchies are one of the most important aspects of business analytics in SAP ERP and SAP BW. The new connectivity improvements will enable support for linked nodes for hierarchies in SAC. Performance improvements by optimizing loading of initial metadata of queries and optimizing handling of metadata of non-input enabled variables, significant improvement in real-time analytics will be enabled by the end of the year 2020.

Enhanced Visualizations

Visualizations enable end users to make rapid decisions by unearthing trends and cycles. With multiple formatting enhancements for visualizations, SAP plans to expedite the decision-making process. Story-level enhancements include changing the model behind the story to enable end users to maintain the story. Multiple new charts as well as an improved chart-building experience will enable end users to rapidly create stories and derive insights from the available data.

Some of the prominent changes expected in the road map are custom fonts, linked analysis for drill levels, and explorer enhancements. Further changes are expected in the functionality of the story builder by introduction of the undo and redo functions.

In this subsection we have learned the enhancements planned for SAC in the road map in terms of enabling real-time decisions. In the next subsection, let us explore the enhancements planned for SAC in terms of developing self-service analytics further.

Analytics Application Development

The real power of the Analytics Application is in the scripting and programmatic functions available for customization. The future versions of the Analytics Designer would have further improvements in terms of the scripting APIs that are supported. Some of the new APIs that would be available are data source refresh API, search to insight API, and data locking API. These new functions would enable the developer to build better applications.

Improved Scripting

The real power of the Analytics Application is in the scripting and programmatic functions available for customization. Future versions of the Analytics Designer would have further improvements in terms of the scripting APIs that are supported. Some of the new APIs that would be available are data source refresh API, search to insight API, and data locking API. These new functions would enable the developer to build better applications.

Additionally, there would be support for new widgets like the Page Book, List Box, navigation panel, and more. New theming enhancements and widget palette enhancements would ensure the new applications are as good as applications built with programming languages.

Finally, there would be better integrations with other components of the SAC platform like BI, planning, and Digital Boardroom.

Mobile App Support

SAC has a robust iOS app and SAP has been rapidly ramping up the Android app. However, there have been limitations with the Analytics Application in terms of support for mobile apps. The future plans for the Analytics Designer would enable mobile app support for both iOS and Android. Also included in future releases are a responsive canvas and containers for developing applications natively built for the mobile app. The responsive screen would ensure the end users get the same experience irrespective of the device types.

Smart Predict Integration

Smart Predict enables end users to build predictive scenarios for data mining and develop predictive models over historical data. SAP has also been rapidly ramping up the Smart Predict portfolio by adding further scenarios in addition to the currently available time series, classification, and regression scenarios. Smart Predict turns out to be a very effective tool for end users and decision-makers make informed decisions around a particular business scenario. However, currently there is no integration between the Analytics Designer and the Smart Predict application although both exist in the same SAC landscape.

In future releases, SAP plans to integrate both the applications and make Smart Predict available in Application Development. This would enable developers to build custom predictive scenarios into custom applications and deliver an end-to-end experience to end users.

1. **Enhancements roadmap for self-service analytics:**

Self-service analytics is one of the primary capabilities of SAC. Enabling the end users to self-analyze data without the need for approaching the IT team. Self-service enables enterprises to arrive at rapid decisions for critical business problems. With industry knowledge the end users are able to contextualize the data and build stories and dashboards for analyzing and presenting the data. If the analytics tool is able to provide tools which offer the end users with a low learning curve to rapidly scale up on the technology, business decisions can be rapidly facilitated. In this section, let us learn the enhancements planned for SAC for self-service analytics.

Augmented Analytics

Augmented analytics with smart features have been the differentiating factor for SAC. With Smart Predict and Smart Insights, SAC supports end users to explore data and build stories around the data. SAC's road map for augmented and smart capabilities include enhanced recommendations for searching, speech to text for NLP communication, and enhanced insights for smart insights. Smart Predict will see the transition from black box AI to explainable AI for increased explainability for time series models. Support for live connectivity will enable the models to work on live data not stored on the SAC landscape.

Bringing in the functionality of natural language processing via Search to Insight will enable executives using Digital Boardroom to converse with the platform in real time without having to even point to a particular data point to explore. Search to Insight is also being enabled for mobile apps enabling better self-service among end users. Live data connectivity support for augmented analytics is also on the road map for SAC.

Explainable Artificial Intelligence experts are needed to understand and interpret not just the results but also the underlying logic behind the same. With these technologies, business decisions involving Artificial Intelligence enable decisions to be more confident and build trust.

SAP recognizes this trend in Artificial Intelligence and machine learning technologies and as part of its road map for SAC has plans to integrate the same. This will enable organizations to embrace AI technologies and be confident of the outcomes and business decisions it supports. This will enable less iterations over training of the AI model and can be sent for production at the earliest with clear understanding of the way the model has delivered results.

Enhanced Modeling

Data preparation is one of the most essential capabilities for analytics. Models enable SAC to bring data across multiple data sources and blend and mash it to create a consolidated dataset. The modeling interface enables end users and analysts to bring strong datasets for data exploration and trend analysis. The datasets can also be consumed via augmented analytics or shared via stories.

A completely revamped modeling experience in terms of Agile BI and enhanced analytics support will enable modeling to be better. The Modeler application is slated for a complete overhaul and will offer the end user with an enhanced intuitive interface for building the model.

Agile BI integrates the concepts of Agile development methodologies into Business Intelligence projects. Agile BI enables continuous development of analytics and KPIs through models and stories. Integrated with processes and continuous deployment methodologies, this can be a very powerful mechanism to reduce the time taken for BI projects development.

To enable Agile BI, models would need to be flexible and editable. Smart Wrangling planned in the next release will enable flexible modeling features. Stories would need to adjustable to changes in back-end models and data changes.

In this subsection we have learned the enhancements planned for SAC in terms of self-service analytics and how they will add business value. In the next subsection, let us learn the enhancements planned for SAC in the planning category.

2. **Enhancements road map for planning**

Planning is one of the core components of SAC. Integrated planning enables end users to combine planning features with Business Intelligence features. Along with core planning features like allocation, spreading, and distribution, SAC provides integrated features like charts, graphs, and tables that can be integrated into a story. Instead of working only with tables, enterprise planners can build dashboards and build in what-if analyses as well as variances and thresholds into the planning process. In this section, let us learn the innovations in planning that are slated to be released for SAC.

Advanced Planning

Advanced planning is enabled in SAC by Data Actions using complex formulas and scripting mechanisms. Disaggregation, spreading, and distribution form the core of planning processes with advanced planning features adding further flexibility in the planning process. The future developments in planning will all be focused toward advanced planning. There are enhancements in data actions and validation rules. Data auditing will enable tracking changes in planning stories, thus enabling users to compare data between versions. A data locking feature will now be available even for disaggregation of values across cells.

Planning Integration: Currently, the interaction between planning functions and predictive functions is limited. However, better integration with planning functions is planned in the future. Planners would be able to leverage segmented time series into the planning grid and enable forecasts on planning data. This would enable improved business planning and enable businesses to form accurate plans with reduced time for reviews and discussions.

Integrations across stories would also enable seamless data access to forecast output that can be consumed over contextual dimensions.

The road map also suggests a single integration between multiple planning platforms like IBP, BPC, and SAC. Additionally, augmented analytics features are also being rapidly made available for planning functions. Smart Predict scenarios would also be enabled for planning, enabling planners to utilize the full functionality of predictive analytics.

As we have understood from above, we have learned the most important enhancements planned for SAC planning including the core planning functions.

5.   **Enhancements road map for a Secure Platform:**

In this section we have learned the most popular trends in analytics where SAC would rapidly have to scale up to enable organizations to be able to make the most of these fast-emerging trends. In the next section, let us learn how to keep track of the release notes for SAC.

Auditing

Auditing is the process of examining of all the activities within any technology landscape. SAC already provides options for tracking data changes and user activities. We have already explored the step-by-step process to monitor data changes in a model as seen in section "Monitoring data changes" in Chapter 9. There are monitoring functions for user activities that are primarily restricted to System Owner and Administrators. For information on a specific data tracking or user activity, the end users would need to reach out to the administrators. This is cumbersome and time consuming.

SAP's future landscape for SAC includes the process of self-service auditing within the system. This will offer further improvement in terms of self-service for end users. The end users would be able to trace back and track changes without having to rely on administrators for the activity. With improved data tracking, the end users would be able to debug issues within the development activities and be able to resolve issues rapidly without reliance on the IT team. The improved auditing will also enable data tracking across the entire life cycle and enable rapid debugging for any changes to the data.

Self-service auditing is also one of the core pillars of Agile BI and wherein the users will be able to build models and create stories in sprints.

Improved Authentication

Authentication is the process of enabling a user to successfully log in to the system. SAC offers multiple mechanisms of logging in including with email and password as well as SSO (Single Sign On). SSO enables users to bypass repeated logging in to multiple systems based on predefined security certificates. Support for Security Assertion Mark-p Language (SAML) along with multiple identity providers (IdP)s are is also part of the road map for security within SAC. SAP has already integrated SAC into the SAP S/4HANA cloud enabling embedded analytics through SAC. This has been possible with the improved use of the SAP Identity Provider. The same concept is expected to be enabled across other applications both on-premise and cloud-enabling seamless integration with SAC.

All of the above will enable users to log in to the system without having to do multiple logins. Moreover, these are proven security protocols and can be part of the entire enterprise security. Custom authentication would also be enabled through the SDKs available including the SDKs for Android and Intune.

Improved Authorization

SAC's robust security model consisting of rights, roles, and users will see further enhancements in terms of providing security across the SAC landscape. This will enable administrators to further tune the security model and provide end users with a simple but secure landscape to operate on. The security enhancements in authorization also include folder-level security. This will enable administrators to ensure only the right people view the right content thus further strengthening the security model. For example, while using a live connection, the authorizations would flow from the back-end system enabling them to be integrated with the authorizations set up in SAC.

Further enhancements in terms of permissions and base-level rights for roles and users would also see new permissions for commenting. A specific security layer for Predictive Scenarios is also expected toward the end of the year, which would enable sharing of specific scenarios with only specific users.

Further enhancements in terms of permissions and base-level rights for roles and users would also see new permissions for commenting. The specific security layer for Predictive Scenarios is also expected toward the end of the year, which would enable sharing of specific scenarios with only specific users.

3. Release plan for SAC:

Over the years, SAC has matured from providing minimal features to a powerhouse with multiple capabilities for Analytics. The three pillars of Business Intelligence, Planning, and Predictive provide enterprises with the core analytical skills needed on a single platform. Being a cloud product enables SAC to be low on maintenance with upgrades and updates being catered to by SAP. SAC continues to be the focus of further innovation in analytics with continued improvements in the planning module. And we have appreciated all of the above in alignment with the road map in the subsections "Future Roadmap for SAC" above. The road map provides the user with information on what to expect in the coming releases. However, the information is limited and provides only a basic overview. For in-depth information on specific releases and features planned for the release, let us navigate to the product update featured on SAC's website. The product updates featured here are in depth and provide complete information on what to expect in the upcoming releases of SAC. Let us learn the step-by-step process to access the release updates.

The step-by-step process is shown in Figure 10-4.

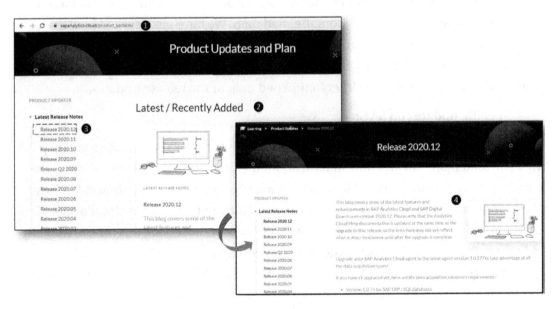

Figure 10-4. *Step-by-Step process to access SAC release updates*

Step 1: Navigate to the URL `https://www.sapanalytics.cloud/product_updates/`. This is shown in Figure 10-4 under ❶. The release notes are arranged as per the release number on the right.

Step 2: The latest release can be accessed directly by clicking on "Latest/Recently Added." This is as shown in ❷.

Step 3: Click on the release number as shown in ❸. The in-depth release information comes up as shown in Figure 10-4.

Step 4: Explore the in-depth information as shown in Figure 10-4 under ❹.

The release notes also enable exploring history of the releases. The road map for SAC provides only the future direction whereas the release updates for SAC provide a complete history as well as information about future releases of SAC. Do note that initially when SAC was in the premature phase, the releases were every two weeks to rapidly accommodate for new features. In the mature phase, the releases are quarterly in tune with the other cloud products of SAP.

In this section, we have explored the release updates of SAC and now understand how release updates are different from the road map. We have also learned how the release updates are different from the road maps and provide additional information.

In the next section, let us learn the process of accessing the specific information about SAC that is not available either in the road map or the release updates.

4. Specific updates on SAC:

In the previous section we learned about release updates as well as the road map for SAC. The road map provides information about the future releases of SAC whereas the release updates provide detailed information a well as release history about SAC. Now let us consider a scenario wherein we would need further information about a specific feature of SAC. SAP provides a robust knowledge base for accessing this information. In addition to the knowledge base, customers can also access the SAP Launchpad to raise an incident for resolving the information. The launchpad can be accessed only with an S user id. Only customer and partners have access to the SAP Launchpad.

The launchpad can be accessed from `http://launchpad.support.sap.com/`. Let us now learn how to access the knowledge base provided for SAC as well as raising an incident for SAC.

The step-by-step process for exploring the knowledge base and incidents through the SAP Launchpad is shown in Figure 10-5.

Figure 10-5. *Step-by-Step Process to access the SAP Support for SAC*

Step 1: From a browser, access the URL `http://launchpad.support.sap.com/`. Enter the S user id and the password. This is shown in Figure 10-5. Note that without an S user id, access to the support site is not possible.

Step 2: The site presents multiple options for accessing the knowledge base or to search incidents. This is shown in Figure 10-5. On selecting the option to search SAP Analytics Cloud in the knowledge base, all the notes as well as articles for specific issues come up.

Step 3: A similar step has to be followed for searching and then creating an incident for a specific issue in the Support Launchpad. Further information for creating an incident in the support launchpad can be found at `https://www.sapanalytics.cloud/resources-support/`. The support site also offers interacting with an expert or scheduling time with an expert for resolving a specific issue in case emergency support is needed.

We have learned in the above step-by-step process how to access the knowledgebase and incidents in the SAP Support Launchpad.

SAP also enables support through its support website https://support.sap.com/en/index.html to enable customers to interact with experts or to report an incident. The support website also serves as an entry point to the support launchpad. The support website enables a one-stop shop for all support activities including the following:

1. Expert chat

2. Scheduling an expert chat

3. Report an incident

4. WhatsApp support

5. View created incidents

The options for the SAP support website are as shown in Figure 10-6.

Incidents

Facing an issue with your SAP software? Search for a solution in our knowledge base. If you are unable to find your answer, contact SAP Product Support using your channel of choice: Expert Chat, Report an Incident or Schedule an Expert.

Expert Chat

Expert Chat in the SAP ONE Support Launchpad instantly connects you to live, technical experts. For even faster access, add the *Expert Chat* tile to your launchpad homepage so you can start the application from there.

Chat with an Expert

Schedule an Expert

Connect with an SAP Support engineer in a live, one-on-one 30-minute call. Tell us your question and the product area that you want to discuss. We'll find an available expert to assist you at a time of your choice.

Schedule an Expert

Report an Incident

Use the SAP ONE Support Launchpad to report an incident for errors related to your products or SAP support applications. An intuitive online form will guide you through the process.

Report an incident

WhatsApp

Subscribe to WhatsApp SAP Product Support channels and receive updates including KBAs, wiki's, guided answers, SAP Notes and "hot tips" on only the products that apply to your needs.

View the WhatsApp subscriptions

View my Incidents

Check the status of incidents that are in progress with SAP or view the resolution of your closed incidents.

View my incidents

View my SAP SuccessFactors incidents

Figure 10-6. Overview of SAP Support website

Let us now learn about the options available in the SAP Support website.

Expert Chat

SAP enables immediate resolution of certain issues by connecting live with an expert in SAC. For a business plan holder, this service is available 24X7. One of the advantages of being able to chat with an expert is the time saved during critical business decisions. Note that for any interaction with SAP, the options for resolving BI issues are to select the component LOD-ANA-BI and for planning issues select LOD-ANA-PL.

Scheduling an Expert Chat

Another option is to schedule an expert chat at a convenient time. This can be done by selecting the component LOD-ANA-BI for Business Intelligence issues and LOD-ANA-PL for planning related issues and selecting the convenient date from the calendar. The schedule appointments require at least three days' notice for the booking.

Report an Incident

Reporting an incident enables users to report issues and raise a specific bug to SAP for SAC. SAP maintains known issues in the form of the knowledge base. However, if there is no satisfactory answer available, a request can be raised, and a support executive will get in touch with the user for resolution of the issue.

WhatsApp Support

As a recent development, SAP has started support over one of the most popular communication platforms. The WhatsApp platform can be used to chat with a support executive to resolve any issue with SAC.

View Created Incidents

This option enables users to follow up on already created incidents. If SAP support has already replied and the status has been changed, this option can be used to check the resolution or to send the incident back to SAP.

In this subsection we have seen the step-by-step process to access specific features for SAC. We have also understood how to create a support incident for SAC.

In this section, we have learned about the release plan for SAC and how to find specific information about releases. We have also learned how to connect to SAP support to access the knowledge base and to connect with a SAC expert. Now we will learn about one more important source of information on the Product Road Map and Future direction for SAC, that is, the Analyst Reports.

The analysts follow the technology company, their customers, as well as partners for a very long period and therefore have access to the leadership teams, get feedback on the product, services, and CIOs and practitioners around the world and therefore offer a valuable source of information on the Product Road Map and Future direction. Analyst reports also provide an unbiased third-party view. Most leading Analysts publish their research and findings on a regular period and include peer comparison charts enable organizations to make the right decision based on their specific needs. For example, an organization running the S/4HANA Enterprise software would benefit most by integrating data analytics with SAC. Analyst reporting bring our nuances with comparisons across specific points within a common group of software applications and their vendors. In the next subsection, let us learn how top analyst firms have covered SAC in their reports.

Customers and partners can also influence SAP product development by submitting a new improvement request at `https://influence.sap.com/sap/ino/#/`

5. Analyst updates on SAC:

As the industry has been steadily accepting SAC as a top contender for cloud-based analytics, it has also found widespread acceptance across the analyst community. Some of the most foremost analysts have covered SAC in their reports, which have been instrumental in the positioning of SAP Analytics Portfolio in the top sections of the analyst reports. Let us understand each of the top analyst reports and how they rank SAC in their reports.

Gartner

Gartner's Magic Quadrant has been one of the most popular reports on evaluating products and applications across the entire IT landscape. The Magic Quadrant for Analytics evaluates the most popular Analytics software vendors across the entire landscape and provides a distinct categorization across four quadrants. SAP has been positioned in the Visionaries quadrant due to the unique concept of SAC being a completely cloud-based product integrating analytics, planning, and predictive technologies. SAP has been steadily rising across the quadrant and with the popularity of SAC and other analytics products gaining popularity, could soon see itself placed in the leaders' quadrant.

The latest Magic Quadrant for Analytics can be accessed from `https://www.gartner.com/en/documents/3980852/magic-quadrant-for-analytics-and-business-intelligence-p`

Forrester

Another very popular report that evaluates analytics software vendors is the Forrester wave. Forrester categorizes the analytics software vendors into waves based on customer reviews, surveys, and features of the product. SAP has been placed as a leader in the latest Wave for Data management for Analytics. Forrester has ranked SAP based on the ability of the data management with SAP data warehousing products including visualization.

The Forrester Wave can be accessed from:

`https://www.forrester.com/report/The+Forrester+Wave+Data +Management+For+Analytics +Q1+2020/-/E-RES157286`

The Forrester Wave for Enterprise BI platforms released in Q3 2019 also names SAP Analytics cloud as a strong performer. The Forrester Report can be accessed from:

`https://www.forrester.com/report/The+Forrester+Wave+Ente rprise+BI+Platforms+Vendor Managed+Q3+2019/-/E-RES151235`

IDC

IDC has recognized the capabilities of SAC in their Marketscape report for Enterprise Performance Management Analytic Applications conducted in 2018. Key findings from the assessment can be summarized as the following:

a. Analytics software buyers are emphasizing on the ease of setting up and administering an easy-to-use landscape.

b. SAP Analytics Cloud provides the best integrated solution for BI, Planning, and Predictive analytics across the enterprise.

The summary of the report with key findings and excerpts can be accessed from this link:

`https://news.sap.com/2018/09/sap-idc-marketscape-worldwide-enterprise-performance-management-analytic-applications/`

BARC

Leading Industry Analyst firm BARC conducted a survey of the top planning applications titled "Planning Survey 19" covering Enterprise software vendors and solution-focused planning products. SAC has been placed as a market leader in the planning applications space with 31 top rankings and 28 leading positions across five different peer groups. Some of the key differentiators noted being the following:

a. Of the respondents, 85% recognized the planning capability of SAC as good and very good.

b. 88% rated the capability of SAC for doing simulations across multiple variables as good and very good.

c. 100% rated the price of the product to the value delivered as good and very good.

The entire survey along with peer comparisons and the parameters measured can be accessed from this link:

```
https://www.sapanalytics.cloud/barc-planning-19/
```

In this subsection we have explored how top analysts rank SAC as a strong player in application for BI and analytics space.

Top analysts which have been covered in this section like Gartner, Forrester, IDC, and BARC are continually tracking top trends in technology spaces. Incorporating deep research with interviews with industry leaders, analyst reports provide a concise but clear view of how trends in technology are shaping up and how software vendors are aligning applications with these trends. Analyst reports form one of the stepping stones for an organization embarking on a Digital Transformation or an application modernization journey. Referring to the Forrester Wave or the Gartner Magic Quadrant enables CIOs and Enterprise Architects to define the right path for long-term business value creation with the right set of tools aligned with organizational goals as well as top trends in technology.

SAC's strong position in the analyst reports and white papers reinforces the recognition of its capabilities and provides a clear view of how the future road map for SAC is aligned with some of the top trends in Analytics.

Summary

In this chapter we have learned industry trends and the how SAC capabilities are aligned to these trends. We have further learned about SAP's vision for the Experience Company driven by the Intelligent Enterprise. We have learned about the following:

1. Top trends in Analytics

2. Industry trends for Analytics

3. The Experience Company driven by the Intelligent Enterprise

4. Future Road Map for SAC

5. Analyst updates on SAC

And that brings us to the end of this chapter and the book. We have learned about the issues faced by ABC Inc. and how implementing SAC will enable ABC Inc. to overcome its current issues and align with the road map to be a fully intelligence-driven organization.

APPENDIX A

The Story Builder

The Story Builder enables building stories from underlying models or directly on datasets. Before we learn the step-by-step process to create a story, let us learn the components of the Story Builder Interface.

The components of the story builder interface are the following:

- Story Tab

- Data Tab

1. Story Tab:

   The story tab allows the story developer access to multiple storytelling components like charts and graphs. The story tab is essential for bringing together data from multiple models and datasets and creating a line of analysis in the form of tables, charts, images, and written text. The story tab is the default tab while creating the story and the components of which are shown in Figure A.1.1.

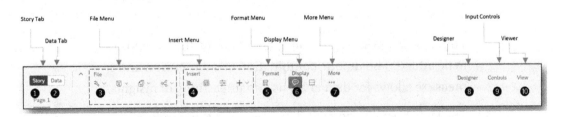

Figure A.1.1. *Story Builder Menu*

© Vinayak Gole, Shreekant Shiralkar 2020
V. Gole and S. Shiralkar, *Empower Decision Makers with SAP Analytics Cloud*,
https://doi.org/10.1007/978-1-4842-6097-5

2.   Data Tab:

Data tab is primarily for data analysts and for a deeper understanding of the data for story builders. The data tab provides end users with the components required for step-by-step data analysis. The components of the data tab are shown in Figure A.1.2.

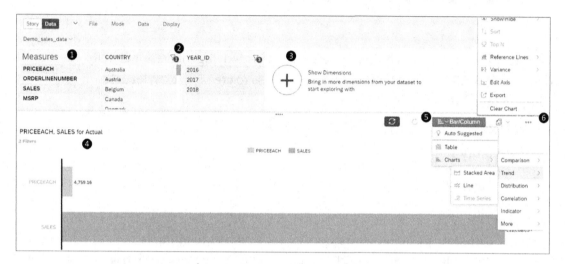

Figure A.1.2. *Data Explorer Menu*

1.   The Measures window ❶ shows the measures available for analysis. One or more measures can be selected for data analysis

2.   The Dimensions windows ❷ show the dimensions added for analysis.

3.   The **Show Dimensions** option ❸ allows for further dimensions to be added. The unique combination of a dimension and a measure allows for data analysis and unearthing of insights.

4.   The Data Visualization window ❹ enables the developer to visualize the data progressively and arrive at the right contextual combination of dimensions and measures delivering the best information. This visualization can then be instantly replicated in a story.

5. The Chart option ❺ enables the story builder to select the most appropriate chart for building the data visualization. This is an important part of the data analysis part since not all charts are appropriate for all types of data analysis.

If the Auto Suggested option is selected, the best option for the data is selected by the SAC engine and displayed accordingly.

6. The three dots at the right end of the window ❻ enable further analysis of the data by bringing in variance display and enabling reference lines. This menu also enables which components need to be displayed and hidden within the chart.

We shall learn about this option in further detail when we discuss the story builder.

3. File Menu:

This menu holds the essential settings for the story file, including options for saving the file and sharing with others. The options incl-uded in the File menu are these:

a) Settings:

This menu brings up the story details, preferences, and query settings. The Settings Menu is shown in Figure A.1.3.

Figure A.1.3. *File Menu Settings*

b) Save Menu:

The save menu brings up options for saving the file, saving as a separate file, and saving as a template. Saving as a template will enable the story as template to be used for other stories. The Save Menu is shown in Figure A.1.4.

Figure A.1.4. *Save Menu*

c) Copy Menu:

The copy menu is useful for copying objects within the same canvas or a new canvas. Multiple options are available within the menu as shown in the figure below. The copy menu is as shown in Figure A.1.5.

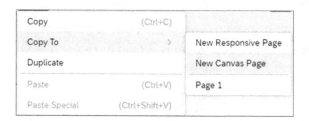

Figure A.1.5. *Copy Menu*

d) Sharing Menu:

This menu allows the story to be shared. On clicking the Share option ❶ as shown in Figure A.1.6, the option to share with a single user or multiple users comes up. The option to also select the level is available as shown in point 2 in the figure below. The Sharing Menu is as shown in Figure A.1.6.

Figure A.1.6. *Share Menu*

4. Insert Menu:

    This is one of the most important menus in the story builder. This menu hosts multiple components that can be used to create a meaningful story, primary components being graphs, tables, and input controls. The components of the menu are as described in Figure A.1.7.

Figure A.1.7. *Insert Menu*

1. **Charts:**

   In any case, this is not part of Insert. Charts are used to represent data visually especially while measuring and comparing KPIs and trends. SAC provides multiple graph types for using in the stories. These can be selected from the Builder Panel, which we will be exploring shortly.

2. **Tables:**

   Tables are representations of data in a tabular format and are used primarily to represent detailed level data.

3. **Input controls:**

   In simple terms, an input control is a filter that is used to restrict data at the story level. An input control can be applied to the story by selecting the dimension or the measure on which the filter is to be applied.

4. **Additional components:**

    SAC provides multiple additional components that can be
    incorporated into a story including Value Driver Trees, which
    we shall learn while discussing about planning as well as
    components of R, which can enable specialized graphs and
    charts.

5. Format Menu:

    This menu has one primary component called Layouts that
    allows for selection of a template from the multiple templates
    provided by SAC or manual addition of custom template as
    shown in Figure A.1.8.

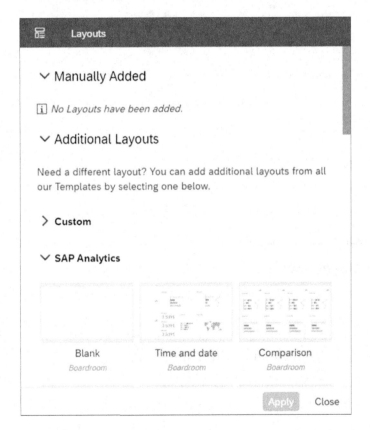

Figure A.1.8. Format Menu

6. Display Menu:

The Display Menu brings up two primary components, that is, the **Comment** mode ❶ and the **Examine** ❷ options. SAC allows collaboration by allowing end users to comment on a particular story or a chart. By enabling the comment mode, these comments are shown, wherever they have been placed ❸. The **Examine** option allows end users to examine the data of a chart in an interface similar to a spreadsheet ❹. Once the examination has been finalized, it can be copied or exported. The components of the display menu are as shown in Figure A.1.9.

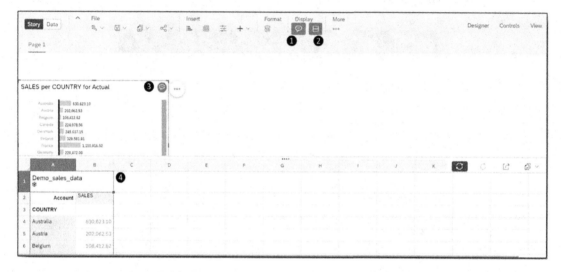

Figure A.1.9. *Display Menu*

7. More:

Multiple additional components are available under the more menu, which are shown in Figure A.1.10.

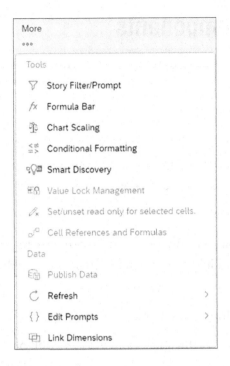

Figure A.1.10. *More options*

8. Designer:

    The designer is the interface where the story can be built
    and designed. It features the canvas where objects including
    charts, tables, and images can be placed. The story builder
    and the styling tabs come up on the right, which can be used
    to enhance and format the story.

9. Input Controls:

    Input controls are interactive components used within
    stories for filtering data. Data components already filtered in
    the particular story can be checked and changed from this
    section.

10. View:

    Once the Story has been created, it can be visualized form the
    View section.

Story Builder Components

The story builder consists of the following components. The Story Builder components are as shown in Figure A.2.1

1. **Data Source:**

   This section contains information about the data source. If a model is being used, the information is shown, which in this particular case is Demo_sales_data.

2. **Add Linked Model:**

   Linking of models is a concept used for bringing together data from multiple models. We shall learn this concept when creating a Story.

3. **Select/Insert Chart:**

   This section displays the charts and graphs available in SAC and allows selection of the most appropriate type for displaying the data. Though new charts are rapidly being added by SAP, the primary charts types available are the following:

   a. **Comparison Charts** that which typically compare values across a dimension. Typically, these are bar graphs.

   b. **Trend Charts** that which compare the value of a measure across a time frame. Typically, these are line graphs.

   c. **Distribution Charts** that show how values are distributed across a dataset. These charts typically compare a measure across multiple dimensions.

   d. **Correlation Charts** that compare values across multiple measures. The purpose of these charts is to determine how one measure can impact another.

   e. **Indicator Charts** that bring out specific values. These could be numeric points to display a single value or bullet charts.

   f. **More Charts** that display how a part compares to the whole of the dataset. Typical examples are donuts or pie charts.

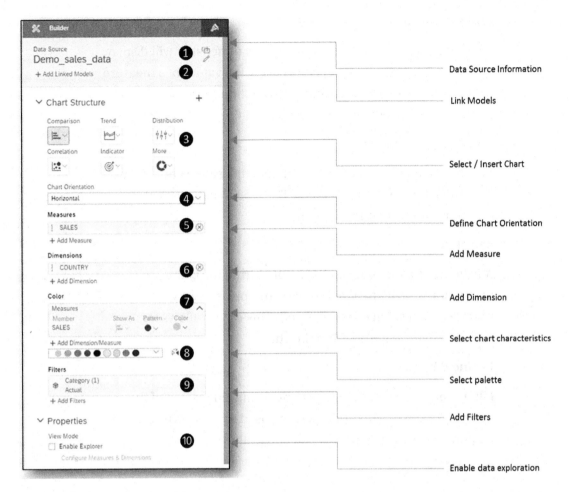

Figure A.2.1. *Story Builder Menu*

4. **Chart orientation:**

   For charts that can be displayed in multiple orientations, there
   are only two variations: vertical or horizontal; this option allows
   for the appropriate selection. For example, a bar graph could be
   vertical or horizontal.

5. **Add measure:**

   The primary component of any data analysis is a data value or
   a measure. This option allows the end user to select the right
   measure for building the story.

313

6. **Add Dimension:**

   A single measure without the context of a dimension will display a single value. For contextually exploring the values of a measure, a dimension needs to be defined. This option allows for the right dimension to be selected.

7. **Chart characteristics:**

   This option allows the end user to define the shape and structure of each measure within the chart. This option can be used to customize the structure of the chart.

8. **Select Palette:**

   A Palette allows for selection of colors within the chart components. SAP provides ready-to-use palettes we well as custom palettes that can be defined. This option allows for selection of the palette within the chart.

9. **Define Filters:**

   Filters enable only certain data to be displayed instead of the complete dataset. This option can be used to select only certain components of the dataset by selecting values.

10. **Data Exploration:**

    This option enables data exploration during a Digital Boardroom presentation. Data exploration can be done also in a story, not necessarily in Digital Boardroom. Data exploration gives the freedom of delving through data as needed and not waiting for answers.

In addition to the Story Builder, the styling panel provides options for formatting each component of the story. The styling tab is as shown in Figure A.2.2. The styling tab consists of the following.

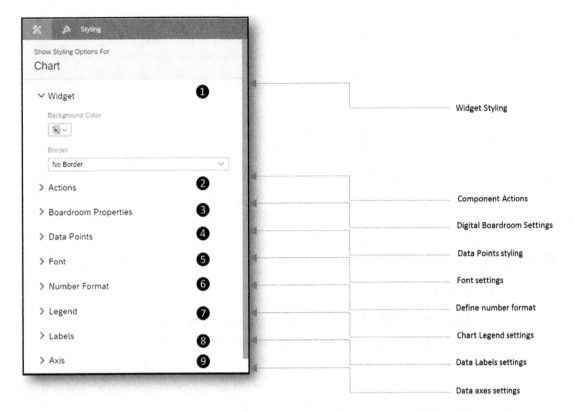

Figure A.2.2. *Story Builder Styling Menu. Note that this would be different based on the version and component.*

1. **Widget:**

   The widget also refers to the block within the canvas. The widget styling options enable the end user to change the background color or to format multiple border options. The Widget styling options are shown in Figure A.2.3.

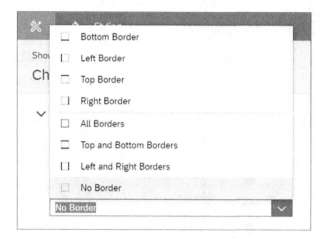

Figure A.2.3. *Widget Menu*

2. **Actions:**

Actions are typically used for changing the order of the widget or blocks as shown in the Figure A.2.4 below:

Figure A.2.4. *Format Actions*

3. **Boardroom Properties:**

Digital Boardroom enables exploration of data over touchscreens directly in boardroom meetings. This option enables the end user to apply the right styling to Digital Boardroom properties. The options for the Digital Boardroom are shown in Figure A.2.5.

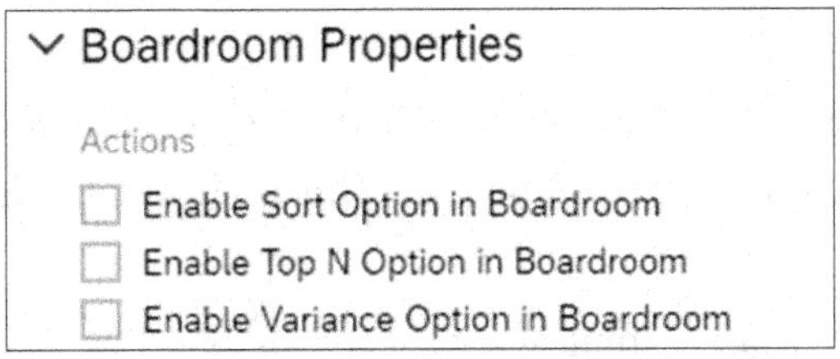

Figure A.2.5. *Boardroom Properties*

4. **Data points:**

Data Points formatting enables us to define how data points in a chart behave. With this feature, you can fill in colors for the data point. Data points styling is shown in Figure A.2.6.

Figure A.2.6. *Data Points*

5. **Font:**

This option defines how each of the fonts within the chart or widget need to be formatted. There are very detailed options for setting up properties of each font. This is shown in Figure A.2.7.

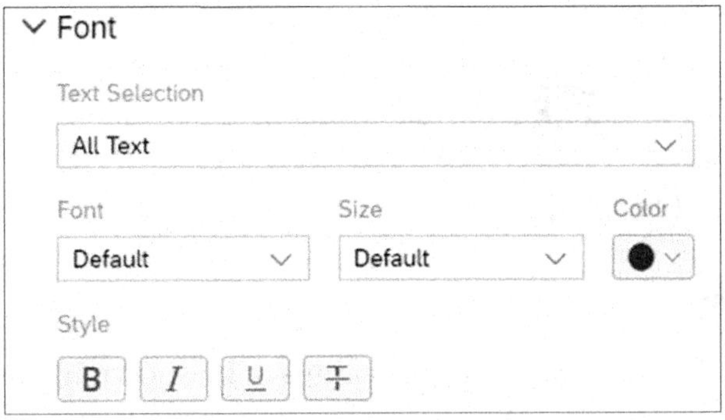

Figure A.2.7. *Font Menu*

6. **Number format:**

    Just like fonts, numerical points can also be defined to match the type of data being displayed. This is shown in Figure A.2.8

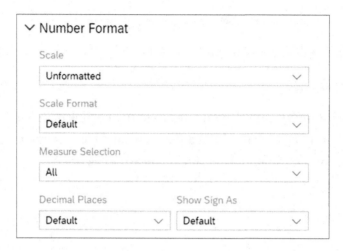

Figure A.2.8. *Number Format*

7. **Legend:**

    Position of the legend and alignment of text can be done as shown below. This is shown in Figure A.2.9.

Figure A.2.9. *Legend Format*

8. **Labels:**

    Data Labels allow numerical values to be shown in addition to the visualization. Formatting options for labels can be done through this option as shown below. This is shown in Figure A.2.10.

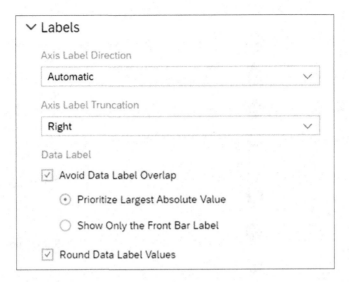

Figure A.2.10. *Label format*

9. **Axis styling:**

The axes on which the data is displayed can be styled through this option as shown in Figure A.2.11.

Figure A.2.11. *Axis Styling*

Chart Properties

Chart properties enable additional features to be enabled for charts and make them more attractive.

Chart Properties

Figure A.3.1 shows the different options available for formatting the charts.

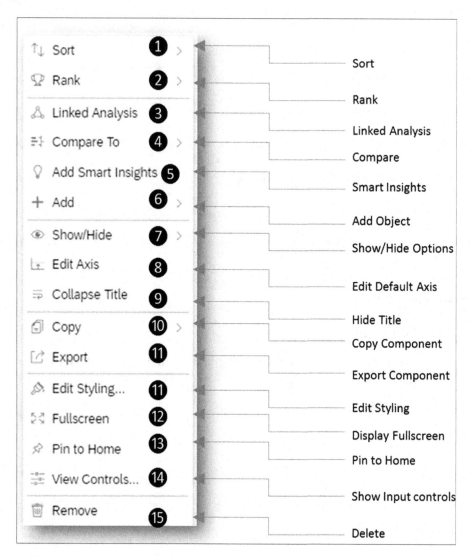

Figure A.3.1. *Chart Properties*

1. **Sort:**

   This function allows the chart to sort the available data. Sort
   options are as below in Figure A.3.2.

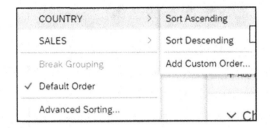

Figure A.3.2. *Sort Options*

2. **Rank:**

   Ranking allows the data to be filtered as per the highest or lowest order. Ranking options are as shown in Figure A.3.3.

Figure A.3.3. *Rank options*

Top N options allow for data to be Ranked on custom user input parameters.

3. **Linked Analysis:**

   Two or more charts can be linked to build a contextual analysis of data. Linked Analysis is a type of advanced filtering wherein data can be manipulated across multiple charts. This is shown in Figure A.3.4.

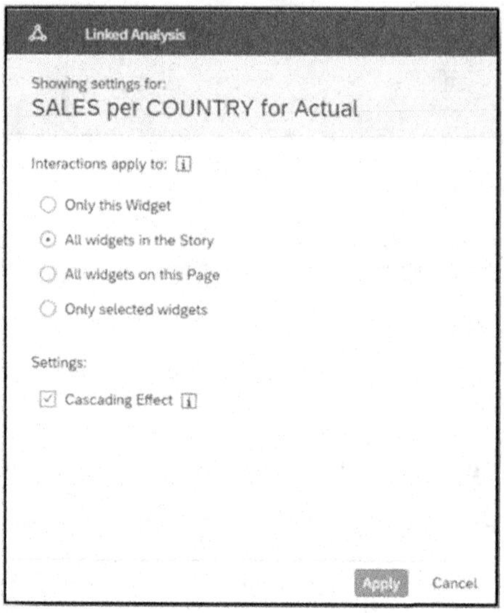

Figure A.3.4. *Linked Analysis*

4. **Compare to:**

   This option allows setting up the variance options comparing one measure to other measures within the story. This is shown in Figure A.3.5.

Figure A.3.5. *Create Variance*

5. **Add Smart Insights:**

   Smart Insights is another feature of the augmented analytics of SAC and adds further value to the already available data. Consider writing a line about this feature here as well. We have learned about this option in further detail in Chapter 5 on Augmented Analytics

6. **Add:**

   This option adds further options that can be added to the chart in addition to the basic information available. These additional options are as below in Figure A.3.6.

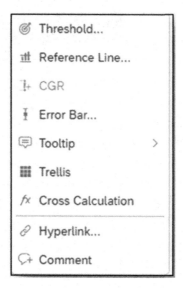

Figure A.3.6. *Add options*

7. **Show/Hide:**

   These are additional formatting options within the chart that allow for aesthetic effect. These options can be enabled or disabled as per the screen real estate or the context of the story. This is shown in Figure A.3.7.

Figure A.3.7. *Show/Hide options*

8. **Edit Axis:**

   As is self-explanatory, this option allows the axes on which the chart is built to be edited and formatted as per requirements.

9. **Collapse Title:**

   In some specific cases, it is required to hide the title of the chart. This can however be made visible by clicking on the arrow. This option allows the title to be collapsed and enabled later accordingly.

10. **Copy:**

    This option allows for the chart to be copied and pasted within the story or a new page. This is shown in Figure A.3.8.

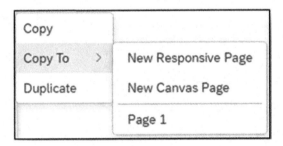

Figure A.3.8. *Options*

11. **Export:**

    Export allows the chart data to be exported as a csv.

12. **Edit Styling:**

    This option allows the styling of the chart to be edited from the Styling panel

13. **Fullscreen:**

    This option allows the chart to be viewed in fullscreen. This option is especially useful while dealing with charts with large quantities of data.

14. **Pin to Home:**

This option allows the chart to be pinned to the Home screen.
This option is useful for end users who must refer to a particular
chart very frequently. Pinning the chart to the home screen allows
the end user to readily refer to the chart without having to explore
the story.

15. **View Controls:**

This option shows the input controls that have been placed on the
chart.

16. **Remove:**

This option deletes the chart from the story canvas.

APPENDIX B

SAC's Analytics Designer

In this section, let us learn the components of SAC's Analytics Designer. To start the Analytics Designer, from the Main Menu, click on Create and then Analytic Application as shown in Figure B.1.1.

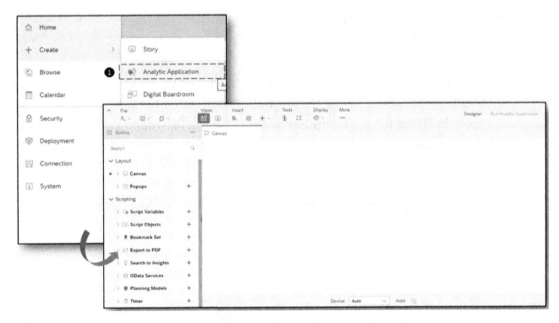

Figure B.1.1. *SAC Analytics Designer*

Let us now learn about the components of the SAC Analytics Designer. We will first learn about the top menu. The top menu contains the components necessary for the analytics application development, saving, and sharing. This menu also holds the toggle menu between the designer and running the application.

The SAC Application Designer menu and its components are as highlighted in Figure B.1.2.

© Vinayak Gole, Shreekant Shiralkar 2020
V. Gole and S. Shiralkar, *Empower Decision Makers with SAP Analytics Cloud,*
https://doi.org/10.1007/978-1-4842-6097-5

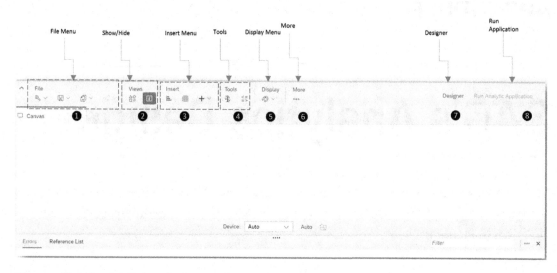

Figure B.1.2. *SAC Analytics Designer Menu*

1. **File Menu:**

    This menu holds the essential settings for the application file, including options for saving the file and sharing with others as shown in Figure B.1.2. The options included in the File menu are the following:

    1. *Settings:*

        The settings menu holds the settings for the application. It has three components as shown in Figure B.1.3.

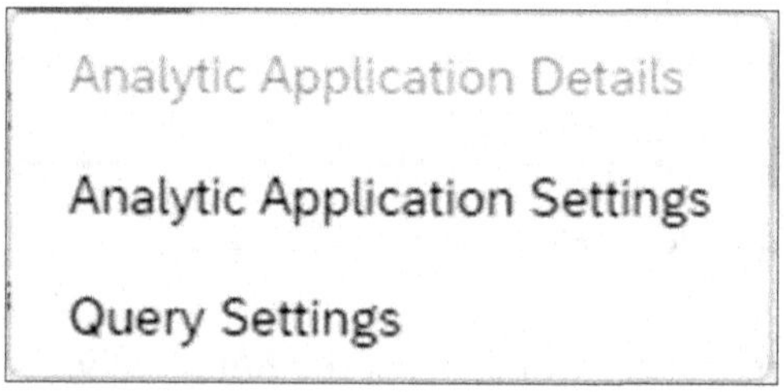

Analytic Application Details

Analytic Application Settings

Query Settings

Figure B.1.3. *SAC Analytics Designer: Settings*

a. *Analytic Application Details:*

Once the application is created, the details are listed under Analytics Application Details. In a new application, the details are grayed out since the details have not been developed yet. This is shown in Figure B.1.3.

b. *Analytic Application Settings:*

As shown in Figure B.1.4, this component of the settings menu consists of two components:

i. **Load invisible widgets on initiation:**

When an application is initiated, it loads multiple components or widgets into the canvas. If there are invisible components in the application, there is little use of loading them since they cannot anyway be seen. However, there might be some components that are crucial to the functioning of the application. This option ensures that these components are loaded when the application initiates. However, there could be an impact on performance due to this selection since now more widgets have to be loaded by the system. This is as shown in Figure B.1.4.

ii. **Load invisible widgets in the background:**

Another option is to load the widgets in the background. The application comes up, but the components or widgets that are not visible are loaded in the background. This enables the application to be loaded rapidly with minimal performance impact. This is shown in Figure B.1.4.

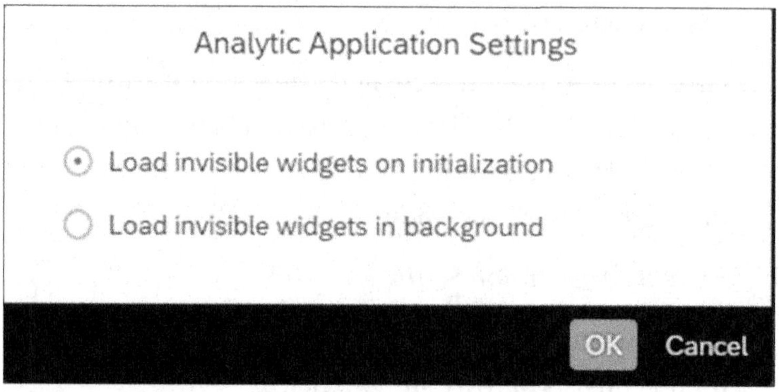

Figure B.1.4. *SAC Analytics Designer: Analytic Application Settings*

c. *Query Settings:*

These settings show how the query can be tuned to enable better performance of the analytic application. The default settings can be changed to enable multiple queries to be run in a batch to reduce the number of database hits, thus improving performance considerably. Also, there is the option to enable Query Merge for SAP BW specific components. This enables faster data fetch especially when the back end is SAP BW. The details of these settings are shown in Figure B.1.5.

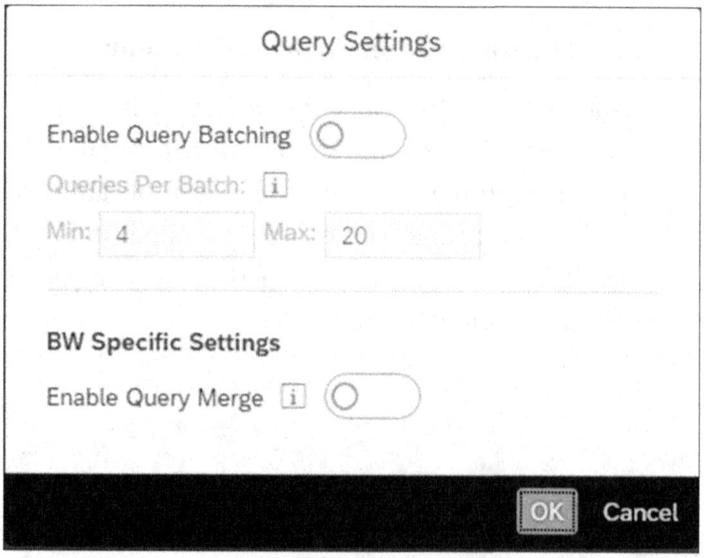

Figure B.1.5. *SAC Analytics Designer: Query Settings*

2. *Save menu:*

    This menu holds the standard options for saving the application. There are options to save the existing application or to save as a new one. This is shown in Figure B.1.6.

Figure B.1.6. *Save Menu*

3. *Copy Menu:*

    This menu holds the options to copy or duplicate components within the application. This is especially useful where an application has similar components that can be tweaked, instead of creating a completely new component from scratch. Components or Widgets can be rapidly duplicated and then options changed. There are multiple options for Special Pasting as well.

    This is shown in Figure B.1.7.

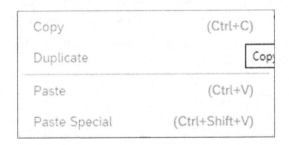

Figure B.1.7. *SAC Analytics Designer: Copy Menu*

4. *Sharing menu:*

Once an application is created, it can be shared across users. This is especially useful for collaboration across multiple users during Agile development of an analytic application. Collaboration is an important aspect of any application and a special feature for SAC. Collaboration allows applications to be shared across specific users and groups. The option to share the Application is shown in Figure B.1.8. We will discuss further on how to share the story in the section "Creating an Analytic Application."

Figure B.1.8. *SAC Analytics Designer: Sharing Settings*

2. **Show/Hide:**

The Toggle Toolbar enables or disables the Outline Menu that holds the components of the Analytic Application.

a. *Show/Hide Outline:*

This option shows the outline panel or the left side panel that holds the widgets that have been used in the application. Individual components when placed on a canvas with scripts and other functions are often termed as widgets. This is as shown in Figure B.1.2.

b. *Show/Hide Info Panel:*

This option shows or hides the information panel at the bottom of the screen. This panel also shows the errors encountered during application development and is very useful during the debugging process. This is shown in Figure B.1.2.

3. **Insert Menu:**

The Insert Menu lists the complete list of components available for building an analytic application as shown in Figure B.1.9. All the components that are required for building an application are included in the list. Typical components in the list are text boxes and filters. Most used components are listed first as shown in ❶, which is a chart and ❷, which is a grid or a table in Figure B.1.9. The other components are listed under '+' which are shown in ❸ in Figure B.1.9. The special components are support for R visualizations, RSS Feeds, and BPC planning sequences. R is a data science language that provides specific visualizations specially designed for data analysis. R code can be included in this component that can be included in the Analytic Application for building a special data visualization.

Similarly, a planning sequence for the back-end SAP BPC or SAP Business planning and consolidation system can be included directly in the application. RSS feeds can be included in the application for up-to0date updates on certain events.

The entire list is shown in Figure B.1.9.

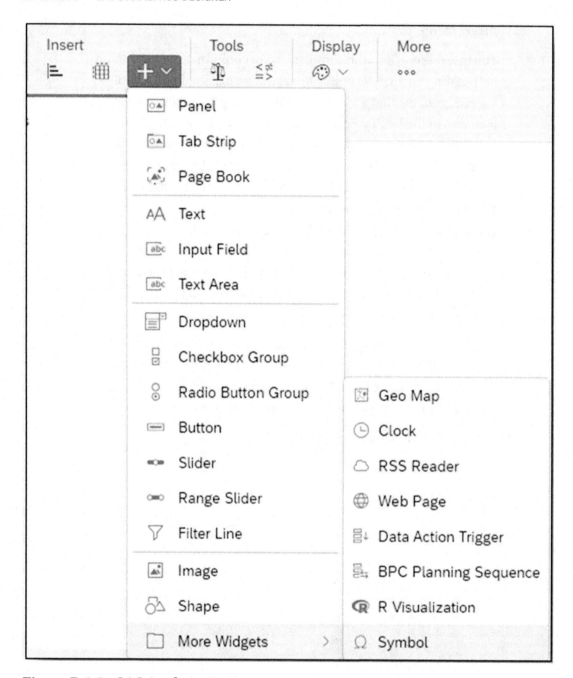

Figure B.1.9. *SAC Analytics Designer: Insert Menu*

4. **Tools menu:**

   The tools menu enables further actions to be performed on the analytics applications. The tools menu is shown in Figure B.1.9.

   a. *Chart Scaling:*

      Chart scaling is especially useful when multiple measures are being compared across the same axes. This helps in comparative analysis of the measures across the same dimensions.

   b. *Conditional Formatting.*

      Conditional formatting is especially useful for formatting objects based on certain conditions. We have learned about conditional formatting in Chapter 4, in the section on "Conditional Formatting."

5. **Display Menu:**

   The options in the Display Menu enable setting multiple options for displaying the components in the canvas. A Display Theme is a collection of settings for each component that is displayed. SAC enables selection from a set of pre-provided themes or the creation of a new theme. This is shown in Figure B.1.10.

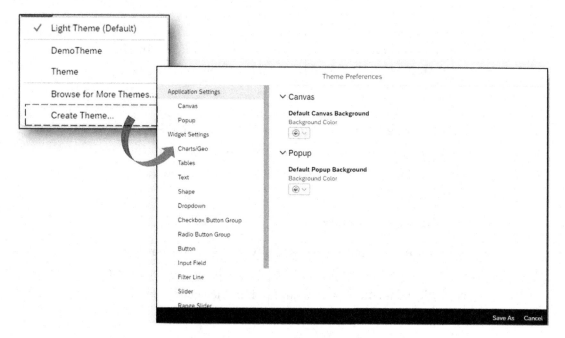

Figure B.1.10. *SAC Analytics Designer: Display Menu*

6. **More:**

   The More Menu consists of components for refreshing the application and editing prompts in the application.

   a. *Refresh:*

   The Refresh Menu brings up the Refresh option within the application. Either the application can be manually refreshed or configured for Auto Refresh based on conditions within this option. This is shown in Figure B.1.11.

Figure B.1.11. *SAC Analytics Designer: Refresh Menu*

b. *Edit Prompts:*

Prompts are inputs to queries on which the application is built. The back end to these data queries is Models. If certain Models require specific inputs, then these are listed under this option. In the model that we have selected, **"demo_sales_sample"**, there are no prompts and hence the option is grayed out.

If multiple models are used to build an application, they would need to be linked. The linking is brought about by linking Variables within the models created. This option enables use of multiple models by linking on variables based on a predefined condition.

These options are as shown in Figure B.1.12.

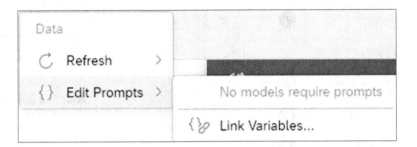

Figure B.1.12. *SAC Analytics Designer: Edit Prompts*

7. **Designer:**

The actual canvas for building the application is called the Designer and can be toggled between building the application and running the application. This is shown in Figure B.1.2.

8. **Run Application:**

This option is used for running the developed application once the development process is completed. This is shown in Figure B.1.2.

We have now learned about the top menu of the Analytics Designer. Now let us learn about the outline menu or the left menu that holds the components for building a custom analytic application.

The Outline Menu consists of components that can be added to the canvas to create an analytic application. The entire Outline Menu is shown in Figure B.1.13. Let us explore each of these components in detail.

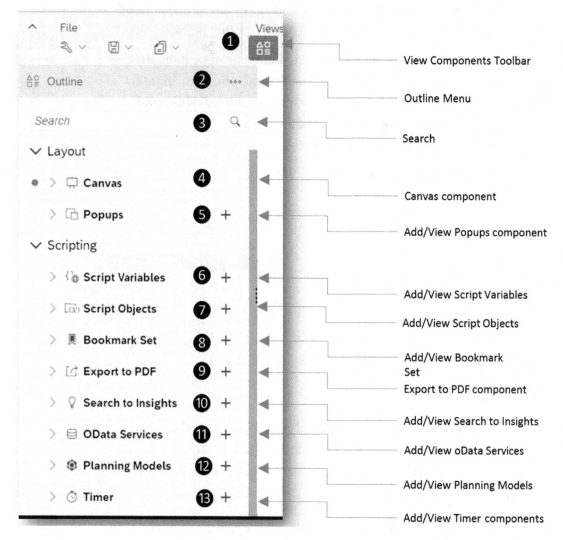

Figure B.1.13. *SAC Analytics Designer: Outline Menu*

1. **View Outline Menu:**

   This button enables the outline Menu from the top menu. Clicking on this option enables toggling the Outline Menu. This is shown in Figure B.1.13.

2. **Outline:**

   Visible components under layout and non-visible components under Scripting are all displayed in the Outline Menu. This is shown in Figure B.1.13 and is useful in locating a component while building and debugging an application.

3. **Search:**

   Search can be used to find components from both the layout and the scripting menu. In case of a complex application, Search provides a quick and easy way to find a component. This is shown in Figure B.1.13.

4. **Canvas components:**

   Canvas components are the widgets that have been placed on the canvas. These are all visible components like buttons, charts, and graphs and other components that can be used while creating an analytic application. These components are grouped under the "Layout" header as shown in Figure B.1.13.

5. **Popup components:**

   Again, a component under "Layout," the Popup components come up in a separate window during certain predefined conditions. An example of a popup is a message that is displayed as a popup. These components are as shown in Figure B.1.13.

6. **Script Variables:**

   A script variable is a reference to a calculation. These can be used as reusable components within the analytics application. Script Variables are as shown in Figure B.1.13.

7. **Script Objects:**

   Script objects act as containers for script functions not linked to any events. They also allow developers to maintain libraries of functions for reusability. Script Objects are as shown in Figure B.1.13.

8. **Bookmarks:**

A bookmark component allows an end user to capture the current state of the application events. A developer can add a bookmark to enable this functionality. In case the end user wants to maintain a reference to the analytic application data at any point in time, bookmarks can be used. This component is shown in Figure B.1.13.

9. **Export to PDF:**

The Export to PDF function allows the developer to add the functionality of exporting the application to a pdf file. This allows end users to download the entire application as a sharable document in portable format. This component is shown in Figure B.1.13.

10. **Search to Insights:**

The Search to Insights component enables the developer to build the entire Search to Insights augmented analytics functionality into the analytics application. This allows end users to communicate with the SAC platform and allow independent data analysis. This component is shown in Figure B.1.13.

11. **OData Services:**

OData or Open Data Protocol enables applications to create and consume query-able RESTful APIs easily into Applications. OData services are extremely useful for connecting to external systems for collecting data. SAC supports the following SAP applications with OData:

- SAP S/4HANA on premise

- SAP BW

- SAP HANA

- SAP BPC

    This component is shown in Figure B.1.13.

12. **Planning Models:**

The Planning Models component allows building planning functions into the analytics applications. Planning applications offer additional functionality over analytics and can be included for building and planning specific analytics applications. This component is as shown in Figure B.1.13.

13. **Timer component:**

Timer components enable adding a timer to the analytic application to trigger events. Typical uses for a timer component would be to schedule, refresh, or send notifications to end users. This component is shown in Figure B.1.13.

Index

© Vinayak Gole, Shreekant Shiralkar 2020
V. Gole and S. Shiralkar, *Empower Decision Makers with SAP Analytics Cloud*,
https://doi.org/10.1007/978-1-4842-6097-5

Printed in the United States
By Bookmasters